Space and Society

Series editor

Douglas A. Vakoch, SETI Institute, Mountain View, CA, USA
    and California Institute of Integral Studies, San Francisco, CA, USA

More information about this series at http://www.springer.com/series/11929

Beth Laura O'Leary · P.J. Capelotti
Editors

Archaeology and Heritage of the Human Movement into Space

 Springer

Editors
Beth Laura O'Leary
Department of Anthropology
New Mexico State University
Las Cruces, NM
USA

P.J. Capelotti
Division of Social Sciences
Abington College
Penn State University
Abington, PA
USA

ISSN 2199-3882
ISBN 978-3-319-07865-6
DOI 10.1007/978-3-319-07866-3

ISSN 2199-3890 (electronic)
ISBN 978-3-319-07866-3   (eBook)

Library of Congress Control Number: 2014946396

Springer Cham Heidelberg New York Dordrecht London

Printed on acid-free paper

Springer is part of Springer Science+Business Media (www.springer.com)

This work is dedicated to Dr. Ben R. Finney, anthropologist of sea and space.

Foreword: Archaeology of Humanity's Space Future

First of all, an early admission!

I am an unabashed child of Sputnik, the Space Race, and my own gravitational attraction to rocket ships that give rise to the exploration of distant worlds.

I have been reporting on humankind's inner passion to have close-encounters with outer space for some 60 years now. Way back then—for 25 cents a pop—I was selling to my more kindly and sympathetic neighbors a handwritten "newspaper" that detailed early rocket flights to the edge of space.

It was apparent that mice, monkeys, and dogs could make the trek. I needed to let them know.

But it was also obvious that getting a true leg up on space, having humans take the journey, was going to be tough-sledding. In fact, in those early years, watching a movie clip of a human zip down a set of rails on a rocket sled at breakneck speed clearly demonstrated that space travel was not for the faint of heart.

Nevertheless, over the following decades I have witnessed and written about amazing exploits, propelled by all manner of human behavior, including sheer inquisitiveness.

From 1957, when the first *artificial* satellite of Earth scooted across the heavens to the epic planting of human footprints on the *natural* satellite of our planet in 1969—undeniably, those milestones of achievement were stunning and remarkable.

However, events like those are precursors lurking in the past. That is, they set in motion humankind's still-to-be-written future in space.

Over the modest increment of time, measured in just decades, human activity in space exploration is societal pyramid building. Space exploration is a result of a collective human enterprise that builds upon centuries of philosophizing, dreaming, scheming ... and then doing.

I do recall my mother advising me in the halcyon days of the Space Race: "The meek shall inherit the Earth ... yes, but the brave ones will go to the moon!"

The ultimate value of vaulting into space is yet to be understood.

To that end, we must preserve, protect, and embrace the astronautical artifacts, inscriptions, and monuments that tell the story, not only for historians of the day but for archaeologists of the tomorrow's yet to come.

This book is an invaluable contribution to better appreciating the evolving field of space archaeology and heritage. It is where sites and artifacts exist in a cultural landscape not just here on Earth, but in outer space and on other celestial bodies—an environment that has been labeled the "spacescape."

The chapters that follow take the reader on a far-flung adventure of what some may tag as archaeology of the future. Sorry to say, we need to think "present tense." We need a legacy to stand on.

For nearly 60 years, tons and tons of handcrafted works of art have been launched into space. The trail of human presence on the Moon still exists. As you read this, telepresence-created wheel tracks on Mars are being made, day after day. Probes are outward bound, chalking up mileage as they speed through the solar system and beyond.

Now is the time to decide what and how to preserve this incredible, indelible record of scientific, technological, and cultural prowess for generations ahead.

For example, Earth's moon is a time-weathered, airless, and dusty place. But it should be considered a cultural and natural landscape in need of preservation.

As an editor of a popular space magazine years ago, I asked the readers a simple question: Why do we call our moon just "the moon"? After all, Jupiter has Europa, Saturn has got Titan. There was need to face facts. There are scads of nifty names out there. Calling our moon "the moon" seems a bit blah.

True, our moon was dubbed Luna by the Romans, then Selene, and Artemis by the Greeks. But when is the last time you heard: "Hey, it's a beautiful Luna out tonight." The response from the name-the-moon competition was overwhelming. Yes, some silly, some off-the-wall, but some profound names were sent in. To me, there was an overall signal: A need for popular possession of Earth's moon. Be it name calling, but so be it.

Consider Apollo 11's Tranquility Base. That's the touchdown spot of the first humans to barnstorm off Earth. At that lunar locale, artifacts from the historic first landing on the moon include discarded life support back packs, tools, and scientific gear. And that's not to mention the first footprints on another world, those of Neil Armstrong and Buzz Aldrin.

Here on Earth we revel at news of "the oldest human footprints" yet discovered. So, why not see the worth of first footsteps left there on the magnificent desolation of a dusty moon?

Part of the answer you'll find within the pages of this book.

Space archaeology and heritage is, in a real sense, fighting for validity. We seem to be in freefall about how to study and value this distinctive aspect of our culture, one that is highly technological.

Secondly, this volume is a clear call for a framework, the needed mechanisms to spark a dialogue to prevent the loss of humankind's early space legacy.

Sputnik, Vanguard, Explorer, Ranger, Surveyor, Lunokhod, Apollo, Viking, Pioneer, Voyager, Spirit, Opportunity, Curiosity, Hayabusa, Chang'e, Chandrayaan, and so many other names, dot the pages of the ever-growing encyclopedia of global space exploration. They are time capsules, testimonials to human knowhow at liftoff.

There are those that brand vagabonds of the vacuum as out of sight, out of mind "space junk." These ground-breaking efforts, though, represent the history-to-date of space investigation, with more to come.

However, without a commitment to preserve our space past, the world stands vulnerable to impacts of the future by many varieties of space travel, be it governmental, commercial, or entrepreneurial.

In closing, let me make a forecast.

I have been writing about space endeavors for many, many years. It seems clear to me that the arduous and costly constraints of accessing space today will not be with us everlasting. The moon will become a tourist stopover. Mars can be permanently populated by the first extraterrestrial nomads off Earth. Enormous solar system distances now charted are likely to fall to ho-hum speed lanes of travel. Even leaps to other stars and their entourage of new worlds will be in sight.

This book is a call to humanity for protection of off-Earth cultural resources. That may well mean preserving Apollo landing sites on the moon as national historic landmarks, as well as regarding far-flung spacecraft as mobile artifacts from space-aged, "prehistoric" times.

In the long run, what's a future for if you can't reconcile the past?

Leonard David

Preface

The idea for this volume was born at the 78th Conference of the Society for American Archaeology (SAA) in Honolulu, HI in 2013. The nine chapters are based on the papers given by a group of scholars who presented at a session entitled "Space and Aviation Heritage: An Archaeological and Historic Preservationist Perspective," co-chaired by Lisa Westwood and Beth O'Leary.

The purpose of *Archaeology and Heritage of the Human Movement into Space* is to provide a creative, thoughtful, and informative overview of a rapidly evolving field that is little more than a decade old. Its range is broad in both theoretical and applied perspectives.

The impetus to publish this collection of papers came from Douglas A. Vakoch, Search for Extraterrestrial Intelligence (SETI), who was involved in creating at the Space and Society Series for Springer International Publishing AG in 2013. Doug has been interested in space archaeology since participating in an earlier symposium on that topic at the Fifth World Archaeological Congress (WAC) in Washington, DC, where one of the editors of this volume, Beth O'Leary presented a paper.

At that first WAC 2003 meeting, several of the "space archaeologists" from different countries met for the first time with the goal to make the broader discipline of archaeology aware of a new sub-field—space archaeology and heritage. It was at this meeting that a resolution was passed by WAC to recognize that "the material culture and places associated with space exploration are significant at individual, local, organizational, national and international levels. As space industries and eventual space colonization develop in the twenty-first century, it is necessary to consider what and how elements of this cultural heritage should be preserved for the benefit of present and future generations" (WAC 2003). A World Archaeological Congress Space Heritage Task Force was created at that time and continues to encourage the investigation and incorporation of space heritage into systems of protection and preservation.

Since 2003, space archaeology and heritage as a discipline has become more inclusive of different approaches, concerns, and goals. There were symposia at the Theoretical Archaeology Group Conference in Sheffield, Great Britain (2005), at the International Council on Monuments and Sites Australia Extreme Heritage Conference (2007), and the sixth WAC conference in Dublin Ireland (2008). As the field expanded, many of us working at universities, museums, and in cultural resource management in the United States, Great Britain, Canada, and Australia, recognized each other as kindred spirits exploring this topic. As several of the archaeologists and preservationists are American, there have now been three separate symposia on space archaeology and heritage at the annual SAA meetings.

The book provides an archaeological perspective to add to the usual scientific outlook of a fascinating series. *Archaeology and Heritage of the Human Movement into Space* addresses issues related to the creation, documentation, preservation, and study of the archaeology of lunar, planetary, and interstellar exploration. It defines the attributes of common human technological expressions within national and, increasingly, commercial exploration efforts, and explores the archaeology of both fixed and mobile artifacts in the galaxy. It presents the research of some of the foremost scholars in the field of space archaeology and heritage. The multiple authors encapsulate various ways of looking at the archaeology of fixed as well as mobile human artifacts in the solar system.

The majority of authors are from the academics, but they also include the chair of the Advisory Council on Historic Preservation and a space engineer. In addition, in the past 5 years there have been several MA theses written in American and European Anthropology departments that concern space heritage; one of the authors of this book is a recent graduate with a space heritage thesis.

Since AD 1957 humans have been creating a vast archaeological assemblage in space and on other celestial bodies. This assemblage of heritage objects and sites attest to the human presence off Earth and, as on Earth, the study of the material remains are best investigated by archaeologists and historic preservationists. As space exploration has passed the half-century mark, it is an appropriate time to reflect on the major events and technological developments of this particularly unique twentieth century area of human history. As missions continue into space and private ventures gear up for tourist visits to space and the Moon, it is also an appropriate time to address questions about the meaning and significance of this cultural material.

Other landmarks in the story of space exploration also contribute to the need for this current overview. The world is approaching the 50-year anniversary of the first lunar landing in 1969. With the recent death of Apollo 11 Astronaut Neil Armstrong and the new national players in space, such as China, and commercial interests, we reflect on the significance of humanity's recent exploration off the Earth. Digital imaging done by NASA's Lunar Reconnaissance Orbiter and twin GRAIL spacecraft have located many Apollo era and earlier spacecraft impact areas and actually found missing Soviet-era Luna spacecraft. This technology is a boon to archaeologists, cultural geographers, and preservationists in locating, recording, and planning for the preservation of significant lunar heritage sites.

Furthermore, NASA has recently (2011) issued recommendations to spacefaring entities that emphasize the need to preserve the historic and scientific values of US lunar artifacts. These guidelines address the issues of how the future of space exploration can incorporate preservation into its missions. In 2013, in the US House of Representatives, there was a bill to establish the Apollo Lunar Landing Sites as a National Historic Park on the Moon. Although legally flawed, it is a significant legislative leap into considering protection of our lunar legacy.

Into this unique environment, *Archaeology and Heritage of the Human Movement into Space* presents a balanced and multidimensional view of material cultural studies. These range from behavioral archaeological to historic preservationist perspectives and focus on why and how we need to interpret and curate cultural remains as a critical part of the complex history of human exploration.

Contents

Contributors

Dr. P.J. Capelotti is a Professor of Anthropology at Penn State University, Abington College. The author or editor of nearly 20 books, his archaeological research has taken him from Indonesia to Svalbard, Franz Josef Land, and the North Pole. He is the author of *The Human Archaeology of Space: Lunar, Planetary and Interstellar Relics of Exploration*. His chapter in this book looks at the discard and ability of space technology to have a life history as they become archaeological objects. He explores both the classical and innovative ways archaeologists can study this class of artifacts.

Leonard David is a space journalist and has been reporting on the space industry for more than five decades. He is former director of research for the National Commission on Space and is co-author of Buzz Aldrin's new book "Mission to Mars—My Vision for Space Exploration" published by National Geographic in 2013.

M. Ann Garrison Darrin is the Managing Executive of Space Department at the Johns Hopkins University Applied Physics Laboratory. She worked in aerospace engineering and assurance technologies which includes electronic components and packaging and material science. With Beth O'Leary she is co-editor of *The Handbook of Space Engineering, Archaeology and Heritage* (2009). She provides the scientific description of the space environment and its effects on human cultural objects and structures in space and on planetary bodies

Milford Wayne Donaldson FAIA is an Architect, appointed by President Obama in 2010 as Chair of the Advisory Council on Historic Preservation. He served as California State Historic Preservation Officer from 2004–2012, working at streamlining Section 106 of the National Historic Preservation Act which led to a national initiative toward sustainability and greening of historic resources. Since 1978, his firm, Architect Milford Wayne Donaldson FAIA Inc., has specialized in historic preservation services. His chapter is a specific case study of Cold War Era properties in California, including structures devoted to research on space. It focusses on their relevance to history and the challenges to preserve them.

Dr. Alice Gorman of the Department of Archaeology Flinders University, Australia is an archaeologist and an adjunct Fellow at the Research School of Astronomy and

Astrophysics at the Australian National University who specializes in the material culture of space exploration with a particular focus on orbital debris or "space junk." In her chapter she discusses orbital debris from a cultural heritage perspective and examines the cultural material related to space exploration and what it can tell us about the history of space exploration and contemporary life on Earth.

Dr. Beth Laura O'Leary provides an overview of the field of space archaeology and heritage and the concept of the cultural landscape of space exploration. Dr. O'Leary is a Professor emerita of Anthropology at New Mexico State University and led one of the first the NASA-funded projects to describe the Apollo 11 lunar landing site at Tranquility Base as an historical archaeological site. She has worked with many of her colleagues toward the protection and preservation of significant sites on the Moon.

Joseph Reynolds is a recent graduate with an MS in Historic Preservation from Clemson University/College of Charleston, South Carolina. His Master's thesis was entitled, "One Small Step: How International Space Law interacts with Historic Preservation." He taught Historic Preservation at Holyoke Community College and is currently a historic preservation consultant in Portland, ME. His chapter examines the unique social and political environment of lunar sites and artifacts in terms of preservation and complexities of legal structures to protect them in the future.

Dr. Hanna M. Szczepanowska is a conservation scientist at the Smithsonian Institution in Washington, DC where she has worked for over 5 years at the National Air and Space Museum studying early satellites, solar panels, thermal protection systems, and heat shields from the Apollo and Space Shuttle missions in the context of museum exhibits. She is the author of *Conservation of Cultural Heritage: Key Principles and Approaches* (Routledge 2013). Her chapter looks at the Space Shuttle orbiters as technologically advanced systems which upon completion of their mission entered the museums as part of humanity's scientific and cultural heritage.

Dr. Justin St. P. Walsh is a Registered Professional Archaeologist who specializes in Greek archaeology, intercultural contact and exchange, and the ethics of cultural heritage management. He teaches at Chapman University in Orange, CA. His forthcoming monograph explores how ancient consumers in the western Mediterranean and trans-Alpine Europe created new identities for themselves through their acquisition of Greek pottery. His chapter looks at the mandates for discard and "design for demise" of space objects in context of other cultural phenomena from cultures such as ancient Athens, Shinto structures in Japan, and others which are purposefully ephemeral in nature.

Lisa D. Westwood is a Registered Professional Archaeologist who has been involved in cultural resources management, contract archaeology, museum curation, and university teaching since 1994. She serves as Cultural Resources Manager at ECORP Consulting Inc., an environmental consulting firm where she specializes in historic archaeology, including Cold War Era aerospace installations in California. She also teaches for California State University-Chico and Butte College. Her chapter looks at international avenues for the preservation of space heritage focusing on UNESCO's World Heritage Convention begun during apex of the Apollo Program.

Chapter 1
"To Boldly Go Where No Man [sic] Has Gone Before:" Approaches in Space Archaeology and Heritage

Beth Laura O'Leary

Abstract This chapter provides an introduction to the field of space archaeology and heritage. It defines archaeology as a sub-discipline of Anthropology which embraces the totality of human experience—it is the study of the relationships between material culture and human behavior. Archaeologists study material culture of all time periods (no temporal limits) and in all places (no spatial limits—it can be done off the Earth, in space and on other celestial bodies). It provides an overview of this recent field and the importance of the concept of the cultural landscape of space exploration.

Introduction

The classic stereotype of an archaeologist is an older man in a pith helmet digging up the remains of the Roman Empire. The modern version is a survey crew, including women, equipped with GPS walking over the surface of federal land in the western US and locating (precisely) prehistoric scatters of artifacts in compliance with preservation law.

Archaeology is, however, not tied solely to prehistoric times or just to the surface of the earth. Archaeology is a sub-discipline of Anthropology which embraces the totality of human experience, in that it should be focused on all times and in all places that humans exist. Specifically, Archaeology is the study of the relationships between material culture and human behavior. It can make substantive contributions to knowledge about human behavior that other disciplines such as history and sociology cannot. Archaeologists can study material culture at all

B.L. O'Leary (✉)
Department of Anthropology, New Mexico State University, Las Cruces, NM, USA
e-mail: boleary@nmsu.edu

© Springer International Publishing Switzerland 2015
B.L. O'Leary and P.J. Capelotti (eds.), *Archaeology and Heritage of the Human Movement into Space*, Space and Society, DOI 10.1007/978-3-319-07866-3_1

times (i.e. the past and the present); there are no temporal limits, except perhaps the future. Present material culture is not off limits to archaeology. Archaeologists can also do investigations in all places. There are no spatial limits. Archaeology has been done on all the continents on Earth and within marine environments. It can also be done on material culture off earth, in orbit, in outer space, and on other celestial bodies like Mars and the Moon.

The foundation of archaeological inquiry into space flight and exploration began with archaeologists who were interested in ship and aircraft wrecks and who proposed they could contain significant data (Gould 1983). Finney (1992) who while researching the technology of the Polynesians who were exploring and settling islands in the Pacific Ocean, suggested it might be worthwhile to think about the space sites (both terrestrial and non-terrestrial) created by both the United States and the Soviet Union.

In 1984, the U.S. National Park Service commissioned a *Man in Space Historic Landmark Theme Study* that inventoried sites in the United States that epitomized the space program in order to include them on the National Register of Historic Places, although none of the proposed sites were actually in space (Butowsky 1984). Launch complex 39 A at Kennedy Space Center in Florida where the Apollo 11 rocket took off for the Moon has importance and exceptional significance because it is directly linked to the first landing site at Tranquility Base on the Moon. In the 1990s William Rathje's interest in garbology led him to publish a paper on the archaeology of "space junk" (Rathje 1999). His interest was in the material culture of the Space era in the form of spacecraft, satellites and debris which orbit the Earth. The term he used at the time to describe this subfield was "exoarchaeology" or the study of artifacts in outer space and those who studied them would be "exoarchaeologists" (Rathje 1999). These terms never really stuck and the sub-discipline is currently identified as space archaeology and heritage.

The definition of space archaeology and heritage is "the archaeological study of material culture found in outer space, that is, exoatmospheric material that is clearly the result of human behavior" (Staski 2009: 19). In a broader sense, it is cultural materials *out there*. Those exoatmospheric artifacts are part of a larger assemblage of materials which until a certain point in time and technological development were confined to Earth but entered the archaeological record some-where else (Staski 2009: 19). In its broadest sense, space archaeology includes all material culture in aerospace and aeronautical realms that relate to the develop-ment and support of all exoatmospheric activities. It encompasses human behavior on the Earth as an anchor to which all space materials are tied.

Space archaeology and heritage posits a cultural landscape of space as a vast connecting network of material remains that originate on earth and incorporate all human materials off Earth. One of its most distant examples is the *Voyager 1* spacecraft which now has reached interstellar space. Launched in 1977, *Voyager 1* is now more than 123 times as far from Earth as the planet is from the sun or approximately 18 billion km away (Lewis 2013: 17). It is humankind's most distant artifact.

The Cultural Landscape of Space

The idea of a cultural landscape as appropriate for space archaeology and heritage starts with the earliest movement of our hominid ancestors out of Africa to colonize virtually all of the earth, and then in modern times (c. AD 1957) a move into space. Unlike terrestrial exploration in order to get humans into space and safely return home they need enormous amounts of technological support. However space is arbitrarily defined in altitude, usually 100 km from the earth's surface, it is a distant and mostly inaccessible place.

Gorman (2005, 2009) has referred to the cultural landscape of space as a "spacescape." The earth is an open system with interchange with the universe: ocean tides are affected by the moon and the weather cycles as influenced by the sun, etc. Earth's inhabitants also send cultural materials into space, some which returns back to Earth. Also celestial material in the form of meteorites, asteroids etc. falls to Earth while some is intentionally collected and brought back, such as the collected 21.7 kg of lunar rock and soil collected by Apollo 11 astronauts Neil Armstrong and Buzz Aldrin (Stooke 2007: 212). The famous photo taken from space by Apollo 8 astronaut James Lovell from their flight above the Moon's surface in 1968 shows Earth as a curious, small, blue marble in a very large universe.

To put the exploration of space into a chronological historical context one can pick several starting places, beginning with objects launched skyward. In the United States in 1783, President John Quincy Adams with Benjamin Franklin watched some of the first launches of balloons, but they did not achieve space; even Chuck Yeager's 1947 flight which broke the speed of sound did not reach space.

The origins of the first uncontested movement of artifacts into space began at the end of World War II when US engaged in a competition to acquire German rockets and German rocket scientists. The V2 rocket became the basis of Cold War missile technology and its descendants launched the first satellites and later propelled humans towards the Moon. The Cold War was played out in space as well as on the surface of the Earth (Gorman and O'Leary 2007: 73). In the early 1950s scientists convened and decided as part of follow up to the International Polar Years (1882 and 1932) to include the study of the whole Earth. The International Geophysical Year (1957–1958) included an international competition for nations to create a satellite which would circle the Earth (Gorman and O'Leary 2007: 76).

The winner clearly was *Sputnik,* which was launched by the USSR into Earth's orbit on October 4, 1957. Several months later *Sputnik 2* ascended into space with a small unfortunate dog named Laika. When space became a technological battleground during this period of the Cold War, physicist Edward Teller reflected fears of many Americans when he said that the US had lost "a battle more important than Pearl Harbor" (Killian 1977: 8).

While *Sputniks* circled the Earth, the US had a series of what were termed 'Flopniks' until finally *Explorer I* went into orbit on January 31, 1958. This was followed by *Explorer II* and then *Vanguard 1,* which successfully launched in March 1958. Although the Sputniks and Explorers deorbited and vanished in fiery flashes, *Vanguard 1* is still in orbit. It is predicted to be in orbit for another 240–600 years. It is currently the oldest human artifact in space. As an artifact it is replete with conflicting and contested properties.

Vanguard

Vanguard is part of the historic context of the Cold War Space Race, an ideological weapon as well as an icon of the scientific exploration of space. As an important object or artifact, it even has its own website—proof of its significance (Vanguard 2013). Even though Vanguard carried no internal scientific instruments, analysis of its orbit revealed that the earth bulged at the equator and was not round but pear-shaped. *Vanguard 1* is essentially preserved in situ even though it is orbiting the Earth every 133 min at an altitude which varies from 400 to 2,400 miles (Oberg 2011). It represents many aspects of the Cold War, as a technological component of the Space Race and symbolic of the competition of manifest destiny between the superpowers. Space was a territory to be won by the most technologically advanced. *Vanguard 1* is a fairly lasting artifact of human meaning and a progenitor in the exploration of space (Gorman and O'Leary 2007: 81). Since that time space has no longer been an empty frontier.

With the advent of the space exploration (c. 56 BP), an exoatmospheric archaeological record was created and is increasing exponentially. Figure 1.1 shows *Vanguard 1,* the 1.5 kg, 15 cm aluminum sphere. Currently NASA tracks objects larger than 10 cm in orbit. It references thousands of satellites, spent rocket stage and breakup debris. Orbital debris has been defined as any human manufactured object in orbit that does not currently serve a purpose and is not anticipated to in

Fig. 1.1 Vanguard 1, launched March 17, 1958, and known as the "Grapefruit Satellite." Vanguard 1 became the second artificial satellite successfully placed in Earth orbit by the United States. It was the first solar-powered satellite; just 152 mm (6 in) in diameter and weighing just 1.4 kg (3 lb) (NASA)

the foreseeable future. Orbital debris can be thought of in some senses as what archaeologists term lithic debitage. Whereas there are pieces and parts of the core discarded (debitage) in the making of a projectile point or stone tool, orbital debris is what is left over or discarded since the original spacecraft mission.

Vanguard 1 became the second artificial satellite successfully placed in Earth orbit by the United States. It was the first solar-powered satellite; just 152 mm (6 in) in diameter and weighing just 1.4 kg (3 lb.) (NASA).

There have been over of 42,000 launches since 1957. Of the over 10,000 + objects tracked by NASA, only 7 % are operational spacecraft, 52 % are decommissioned satellites, upper stages and mission related object and 41 % are debris from the fragmentation of orbital objects (Gorman 2009: 382). Humans have also launched spacecraft that have flown towards and past many celestial bodies. Eight bodies have been orbited and seven (Earth, the Moon, Venus, Mars, two asteroids 433 Eros and 25143 Itokawa, and Saturn's Moon Titan) have had landers along with other material culture on their surfaces. Other missions have collided with each other and/or celestial bodies (Gold 2009: 309). These objects are of interest to both scientists investigating changes to the objects' physical states or "space weathering" due to the environments in which they reside, and to archaeologists as heritage objects worthy of study and preservation.

Artifacts on Celestial Bodies

While space objects are mainly technological in nature, some objects actually on other celestial bodies are solely symbolic and serve as mementos and memorials. These items include a mission patch from Apollo 1 which commemorates the astronauts who died in a tragic fire at Cape Canaveral, and which was left on the lunar surface by the Apollo 11 astronauts at Tranquility Base. A gold olive branch and a silicon disk etched with goodwill messages from 73 of the world's nations, without a message from the Soviet Union, was also left at Tranquility Base (Lunar Legacy website 2013). In 1969 the Apollo 11 astronauts planted an American flag (with some resistance from the regolith) on the lunar surface as one of the first activities which they engaged in after landing (Chaiken 1994: 316). This behavior and the flag are symbolic of claiming territory and victory set by historical precedent. To many nations, even those involved in the effort, that American flag on the Moon was an impudent and proprietary gesture. To the Soviets, it must have signaled a defeat in space (O'Leary 2009a: 771; see Fig. 1.2).

The Moon today has over 100 metric tons of cultural materials from several nations, most of it clustered near the lunar equator. One of the most recent sites, created on December 14, 2013, was the successful landing of China's *Chang'e* spacecraft and on the Moon in the Sinus Iridum, a plain of basaltic lava called the Bay of Rainbows. Its *Yutu's* mission was to explore the moon for several months. China has become the third nation to soft land robotic hardware on the Moon (David 2013).

Also there was an archaeological site which created a crater with cultural debris from the impact at the lunar south pole by the LCROSS mission in 2009 (LCROSS

Fig. 1.2 In 2009, NASA's Lunar Reconnaissance Orbiter passed over the Apollo 11 landing site, providing a new view of the historic spot. The astronaut path to the TV camera is visible, and one can identify two parts of the Early Apollo Science Experiments Package (EASEP), the Lunar Ranging Retro Reflector (LRRR) and the Passive Seismic Experiment (PSE). Astronaut Neil Armstrong's tracks to Little West crater (33 m diameter) are also visible (*unlabeled arrow*) (NASA/Goddard Space Flight Center/Arizona State University)

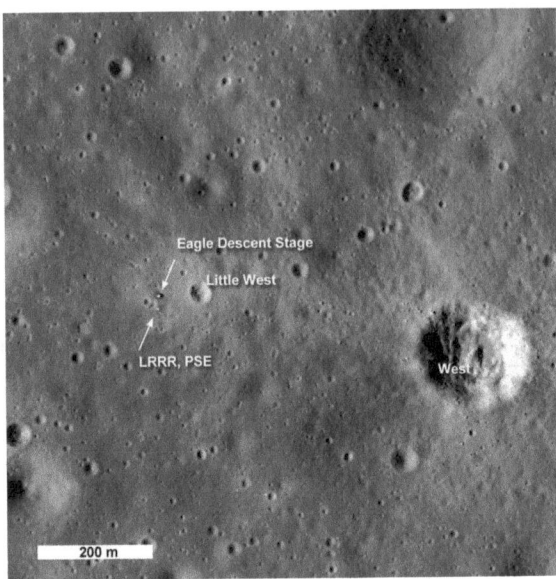

2013). Many objects placed on the moon, of course, usually have a scientific purpose, like several of the lunar laser ranging retro-reflectors left by the Apollo Program and which have gathered data since 1969 and continue to do so at many observatories, while others are more symbolic of political and social concerns. The first probe ever to impact the lunar surface is the USSR's *Luna 2* in September 1959. It delivered a Soviet pennant to the Moon (Stooke 2007: 15). In July 1999 the US *Lunar Prospector,* one of NASA's missions to map the Moon, was deliberately crashed into a lunar crater and carried a small vessel with some of the cremains of the astronomer Eugene Shoemaker, after whom the crater was renamed (Stooke 2007: 394).

Capelotti (2010: 25) has dubbed the entire archaeological assemblage from the Apollo program as the "Apollo culture." This is based on the standard use of archaeological terminology for cultural phases and areas. Without returning to archaeologically survey the lunar sites, the historical documents and visual media prevalent at the time, along with recent digital imagery provide excellent evidence of material remains that epitomize an Apollo era of space exploration (1963–1972). This assemblage both on the Moon, in prototype and those returned to Earth (e.g. Apollo 11 Command module), provide evidence of what Schiffer (1999: 2) would call the "ceaseless and varied interaction among people and various kinds of things."

The Course of Space Archaeology

The direction of space archaeology follows the trajectory of industrial archaeology where analyses are being done on a unique and extraordinary part of the evolution of human technology (Gorman and O'Leary 2013: 419). Olsen (2003: 100) has argued for paying attention to and studying the opposite of how the subject created the object, with a focus on the narrative that everything is language, action, mind and human bodies, and look instead at how the objects construct the subject. For all the recording of recent past there is a surprisingly incomplete bricolage of records for space endeavors. In great part, space artifacts and features have missed our archaeological gaze (Gorman and O'Leary 2013: 419).

The first "archaeology" done in space happened in 1969. The life history of outer space objects was undertaken by Apollo 12 Astronauts Alan Bean and Pete Conrad when their spacecraft landed near the Surveyor 3 robotic spacecraft (O'Leary 2009b: 30). The Surveyor 3 site was photographed, the location was mapped and several samples were collected from the spacecraft (see Fig. 4.2). Although not slated to be an archaeological investigation, Capelotti (1996: 153) has called this "the first example of extraterrestrial archaeology and—perhaps more significant for the history of the discipline—formational archaeology, the study of environmental and cultural forces upon the life history of human artifacts in space."

Also it should be remembered that most of the landscape of space off Earth contains no evidence of actual human physical presence. There are few places human beings actually trod (i.e. the Moon); more common are places where humanity has have sent its robots. As recently as today a robot provides information on the landscape of Mars and links people to that place remotely. Some space scientists doubt humans will ever land again on the Moon or on to Mars in the next 50 years. The cost and logistics appear to be too great. In the years following the utilization of the Space Station *Mir* (now demolished and lying in the Pacific Ocean) and the counting all the crews of the International Space Station few human have ever gone to space, but our material culture in the form of satellites, probes and robotics has increased its presence exponentially in space and on other celestial bodies since exploration began.

A critical time and place in the cultural landscape of space is the site at Tranquility Base, created by Apollo 11 astronauts for over 21 h on July 20, 1969. The features include the trails made by Astronauts Armstrong and Aldrin and a variety of scientific tools and debris from their exploration, including cast off overshoes.

Although this site is within the continuum of exploration it is an example of extremely remote heritage. It is a single component created and occupied by only two people doing fairly scripted behavior at an event which had never happened before. The site was to a large extent self–documenting. The period from the earliest touchdown on the Moon to just before takeoff from the Moon is documented more completely than any other historic event in the twentieth century (O'Leary 2009a: 762). It lies on the outer edge of human exploration even when compared to polar exploration sites in the arctic regions and on Antarctica. One important task for archaeologists is to compare how space exploration sites are different or similar to other past sites.

Since the Lunar Reconnaissance Orbiter on the 2009 LCROSS mission flew over the lunar surface to more accurately map the Moon, we have an

Fig. 1.3 A view from the
Lunar Reconnaissance
Orbiter of the landing area
of Apollo 17 in the region
of Taurus-Littrow. The LRO
had earlier maneuvered into
a 50-km mapping orbit,
allowing it to gain closer
images of archaeological
sites on the Moon than
any previous orbiter
(NASA/Goddard Space
Flight Center/Arizona State
University)

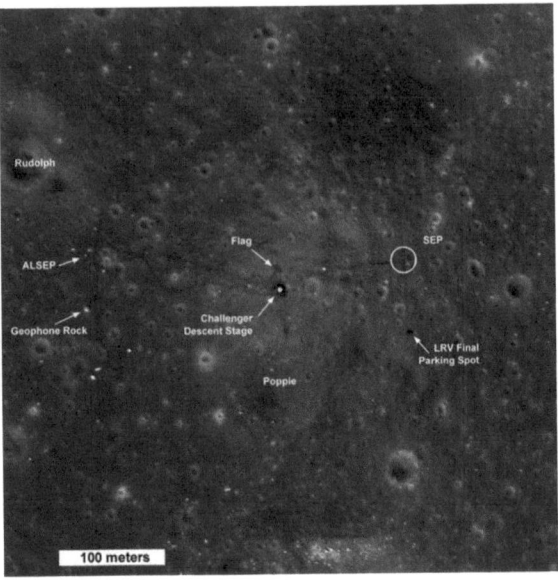

unprecedented opportunity as archaeologists to look remotely at sites on the Moon
untouched by humans since 1972 and investigate the natural transformations to the
material culture left there. Figure 1.2 is an example of the Apollo 11 site as it was
digitally imaged by the LRO in 2009. Figure 1.3 is the last Apollo mission, Apollo
17 as imaged by the LRO. The trails there were made both by humans walking and
by being carried by a lunar rover on Apollo 17 that was abandoned after use and
is still on the Moon's surface. The technology that allows for precise images and
locational information created by scientists to investigate the Moon, other celestial
bodies and space in general can provide archaeologists with the tools to do remote
investigations of cultural material and sites in space.

The academic discipline of archaeology includes its applied field of Cultural
Resource Management, which has as its focus compliance with national and international preservation law and regulation. The archaeological record is only as
complete as what remains undisturbed and/or protected from cultural and environmental disturbance. Currently space sites and objects exist in a kind of legal void.
What is in outer space and on other celestial bodies is governed by a complex tan-
gle of legal authority at the state, national, and international level.

In the U.S. preservation attempts at the individual state level have taken
place, both in California and New Mexico. In 2010 the artifacts and structures
(although not the actual lunar surface which no one owns) at the Apollo 11 lunar
landing site were placed on those respective state registers of cultural properties
(O'Leary et al. 2010; Westwood et al. 2010). The idea was to link the relevance
of each state's history and involvement in the space program to those things left
on the Moon at the iconic first site of lunar exploration. These efforts were largely

symbolic intended to raise public consciousness and to influence decisions to make the site a US National Historic Landmark and eventually to be placed on the World Heritage List.

Few archaeologists or people in general would deny the significance of those artifacts on the Moon, but the public also acknowledges their worth by an increasing demand to own historic space objects. With the recent death of Apollo 11 astronaut Neil Armstrong, who was the first to step on the moon, there is an increased awareness of the "historic" nature of his behavior and an appreciation in the value (both historic and economic) of space artifacts.

As objects and sites reach the golden 50 year mark, like wedding anniversaries, something has to be done to mark and recommit the relationship among the partners. Even at the beginning of the Space Race, replete with the tensions of the Cold War which was, in part, played out in space, the world became enamored and entranced by things "spacey." Music of that era included homage to the early satellite Telstar, and to drink Tang was to be like an astronaut. The Swedish rock group the *Spotniks* chose their name in 1961 as a play on the Soviet satellite, and still tour today. Moon rocks, while studied by scientists, also became prestigious gifts to nations and museums around the world. The internet today provides a global marketplace for anything from a chunk of a meteorite to something flown in space. One example is an object from Apollo 16 (a small pin flag used on the Moon as part of the personal gear of one of the Apollo astronauts) that was auctioned in January 2010 for $16,000 USD.

Starbucks on the Moon

What is the future for lunar preservation? In 2011, NASA issued recommendations to space faring entities on how to protect and preserve the historic and scientific value of U.S. Government artifacts when other commercial enterprises or nations return to the Moon (NASA 2011). NASA (2011) recommends a higher level of protection for Apollo 11 and 17 than the other US sites on the Moon and they make no recommendations for lunar sites belonging to other countries.

The U.S. government, according to the Outer Space Treaty (OST) ratified in 1967, still owns the artifacts placed on the Moon, while prohibiting the ownership of the Moon or other celestial bodies and emphasizing the importance of access to space and international cooperation (Hertzfeld and Pace 2013: 1049). However, the OST does not address preservation issues of significant sites at a time when there are new players such as China and private entities such as Google, who is supporting a competition called the Google Lunar X prize where groups compete to be the first commercial venture to place robotics on the Moon.

The NASA (2011) recommendations are only guidelines that essentially seek ways to avoid damage to US property on the lunar surface. They are very important in that they recognize that future lunar visits could damage and destroy the historic and scientific values of artifacts and sites from the earlier space age.

On July 8, 2013 two U.S. Congresswomen introduced into the US Congress H.R. 2617, The Lunar Landing Legacy Act which proposed to designate the Apollo Programs landing sites and artifacts as a U.S. National Park under the U.S. Department of Interior (U.S. Congress 2013). This action follows an earlier attempt in 2000 to designate the Apollo 11 lunar landing site as a National Historic Landmark which was not supported by NASA or the U.S. Keeper of the National Register of Historic Places (O'Leary 2009a: 763) H.R. 2617 has been deemed legally flawed because it fails to address interests of other nations that have visited or will visit the Moon and it is perceived as a unilateral U.S. action to control parts of the Moon (Hertzfeld and Pace 2013: 1049). In effect, it can be perceived as a claim of sovereignty over the lunar surface.

H.R. 2617 has not been passed but it does show Congressional interest which acknowledges the importance of preserving and providing safeguards for its property, outside of simply declaring "ownership" of the American cultural material on the Moon. The treatment of these sites as well as the early *Luna* sites on the Moon put there by the former Soviet Union should not be based solely on what is owned but what humanity as a whole thinks should be protected from damage and destruction because the sites represent a series of extraordinary events that brought the human race for the first time to another celestial body. The idea of multilateral or international agreements, like the treaties for Antarctica, among the major players with material on the Moon: the U.S., Russia and China is ranked as superior to H.R. 2617 (Hertzfeld and Pace 2013: 1050). There is certainly the necessity to build cooperation in recognizing each nation's interest in the lunar sites and artifacts. An international agreement would reinforce the basic principles of existing space treaties and may be more expedient than amending the OST.

What is valuable for scientific investigation, archaeological investigation and the public who wants to buy space objects and at some time visit the moon as tourists? Today's robotic X Prize competition by the Google affords the opportunity for commercial space exploration, and other nations such as China are currently involved in space. As in remote places like Antarctica, when it became available for tourism, the potential for the damage and destruction of sites increased.

If commercial development follows the colonization of all parts of the world, it certainly will be the same for space. There may one day be a Starbucks on the Moon if it follows the same trends of tourism on Earth at those previously inaccessible places like the Forbidden city in China, war torn zones in Afghanistan, and late 19th and early 20th century polar exploration camps on Antarctica. In the absence of a strong international legal framework for space preservation and heritage, we are bound to lose significant objects and sites.

Schiffer (2013: 177) has argued that although archaeologists have been unable to visit sites off earth there is an extraordinary amount of documentation. Capelotti's (2009, 2010) catalogue of archaeological remains on the Moon and other celestial bodies is one model to investigate the ways we have gone into this extreme environment. Capelotti (2004) has deemed space as the final archaeological frontier. It is the place now to be explored archaeologically.

Conclusions

Reflecting on the nature of space exploration in the last 50 plus years, it is important to remember that the heavens and celestial bodies have always been the property of the world's peoples. Only in the western scientific thought are the heavens usually separated from the Earth. To give two examples: traditional Navajo narratives do not separate Mother Earth from Father Sky; for the Southern Tutchone of the Yukon Canada, the creator Crow put the Moon in the sky for both people and animals. Most indigenous and pre-modern worldviews see the heavens and earth as one system. Archaeology on Earth was linked to outer space and other celestial bodies as soon as humanity learned how to explore that realm. Humans are linked to the heavens now because, with great effort, we can get there. Humans can boldly go where no humans have gone before but the recognition of the earth as a place within the solar system also entails a responsibility for what we can agree upon as a species is significant to our heritage. It is time to acknowledge space archaeology and heritage as vital discipline as the field moves towards the future. Humanity must find ways for space objects to be considered for protection and how they can be preserved for future generations in order to understand their scientific, technical and social significance.

References

Butowsky, H. A. (1984). *Man in space: a national historic landmark theme study*. Washington, DC: US National Park Service, U.S. Department of Interior.

Capelotti, P. J. (2010). *The human archaeology of space: Lunar: planetary and interstellar relics of exploration*. Jefferson, North Carolina: McFarland & Company Inc.

Capelotti, P. J. (2009). The culture of apollo: A catalog of manned exploration of the moon. In A. Darrin & B. O'Leary (Eds.), *Handbook of space engineering, archaeology and heritage* (pp. 421–441). Boca Raton: CRC Taylor and Francis Press.

Capelotti, P. J. (2004). Space: The final [archaeological] frontier. *Archaeology, 57*(2), 108–112.

Capelotti, P. J. (1996). *A Conceptual Model for Aerospace Archaeology*. Rutgers University, Ph.D. Dissertation, UMI: 9633681.

Chaiken, A. (1994). *A man on the moon: One giant leap*. Alexandria, VA: Time-Life Books.

David, L. (2013). China lands on the moon: Historic robotic lunar landing includes first Chinese rover. Accessed December 20, 2013. http://www.space.com/23968-china-moon-rover-historic-lunar-landing.html

Finney, B. R. (1992). *From sea to space*. Palmerston North, New Zealand: Massey University.

Gold, R. (2009). Spacecraft and objects left on planetary surfaces. In A. Darrin & B. O'Leary (Eds.), *Handbook of space engineering, archaeology and heritage* (pp. 399–419). Boca Raton: CRC Taylor and Francis Press.

Gorman, A. C. (2009). Cultural landscape of space. In A. Darrin & B. O'Leary (Eds.), *Handbook of space engineering, archaeology and heritage* (pp. 335–346). Boca Raton: CRC Taylor and Francis Press.

Gorman, A. C. (2005). The cultural landscape of interplanetary space. *Journal of Social Archaeology, 5*(1), 85–107.

Gorman, A. C., & O'Leary, B. (2007). An ideological vacuum: The cold war in outer space. In J. Schofield & W. Cocroft (Eds.), *A fearsome heritage: Diverse legacies of the cold war* (pp. 73–92). Walnut Creek, CA: Left Coast Press.

Gorman, A. C., & O'Leary, B. (2013). The archaeology of space exploration. In P. Graves-Brown, R. Harrison, & A. Piccini (Eds.), *The oxford handbook of the archaeology of the contemporary world* (pp. 409–424). Oxford: Oxford University Press.

Gould, R. A. (Ed.). (1983). *Shipwreck archaeology*. Albuquerque, NM: University of New Mexico Press.

Hertzfeld, H. R., & Pace, S. N. (2013). International cooperation on human lunar heritage. *Science, 342*(29), 1049–1050.

Killian, J. R. (1977). *Sputnik, scientists and eisenhower: a memoir of the first special assistant to the president for science and technology*. Cambridge: MIT Press.

LCROSS Mission NASA-LCROSS. Accessed December 20, 2013. http://www.nasa.gov/mission_pages/LCROSS/main/index.html

Lewis, T. (2013). Voyager enters unchartered territory. *ScienceNews*: 17.

Lunar Legacy Website. Accessed December 20, 2013. http://spacegrant.nmsu.edu/lunarlegacies

NASA. (2011). *NASA's Recommendations to Space-Faring Entities on How to protect and preserve the historic and scientific value of U.S. Government Artifacts*, July 20, 2011. Accessed December 20, 2013. http://www.nasa.gov/directorates/heo/library/reports/lunar-artifacts.html

NASA. (2013). NASA-LCROSS. Accessed December 20, 2013. http://www.nasa.gov/mission_pages/LCROSS/main/index.html

Oberg, J. (2011). Satellite turns 50 years old … in orbit. NBC News, March 17, 2008. Accessed 20 December 2013. http://www.nbcnews.com/id/23639980/

O'Leary, B. (2009a). One giant leap: Preserving cultural resources on the moon, In A. Darrin & B. O'Leary (Eds.), *Handbook of Space Engineering, Archaeology and Heritage* (p. 780). Boca Raton: CRC Taylor and Francis Press.

O'Leary, B. L. (2009b). Evolution of space archaeology and heritage. In A. Darrin & B. O'Leary (Eds.), *Handbook of space engineering, archaeology and heritage* (pp. 29–47). Boca Raton: CRC Taylor and Francis Press.

O'Leary, B.L, Bliss, S., Debry, R., Gibson, R., Punke, M., Sam, D., Slocum, R., Vela, J., Versluis, J., & Westwood, L. (2010). The artifacts and structures at tranquility base nomination to the New Mexico state register of cultural properties. Accepted by unanimous vote by the New Mexico Cultural Properties Review Committee on 10 April 2010

Olsen, B. (2003). Material culture after text: Re-membering things. *Norwegian Archaeological Review, 36*(2), 87–104.

Rathje, W. (1999). An Archaeology of Space Garbage, *Discovering Archaeology* (October): 108–122

Schiffer, M. B. (1999). *The material life of human beings, artifacts*. London/New York Routledge: Behavioral Communication.

Schiffer, M. B. (2013). Archaeology of the space age. *Archaeology of Science, 9*, 163–183.

Staski, E. (2009). Archaeology: The basics. In A. Darrin & B. O'Leary (Eds.), *Handbook of space engineering, archaeology and heritage*. Boca Raton: CRC Taylor and Francis Press.

Stooke, P. J. (2007). *The international atlas of luna exploration*. Cambridge: Cambridge University Press.

U.S. Congress. House. (2013). A Bill to Establish the Apollo Lunar Landing Sites National Historical Park on the Moon, and for other purposes (HR 2617 July 8, 2013). Accessed December 20, 2013. http://thomas.loc.gov/cgi-bin/query/z?c113:H.R.2617

Vanguard. (2013). Live real time satellite tracking and predictions: *Vanguard 1*. Accessed December 20, 2013. http://www.n2yo.com/?s=5

Westwood, L., G. Gibson, B. O'Leary, J. Versluis. (2010). Nomination of the Objects associated with Tranquility Base to the California State Historical Resources Commission. Accepted by unanimous vote to the California State Register of Historical Resources on 30 January 2010

Chapter 2
The Impact of the Space Environment on Material Remains

M. Ann Garrison Darrin

Abstract This chapter provides the scientific introduction to environments which are exo-atmospheric and removed from Earth's protective blanket (our own magnetosphere) and the effects of space on human material culture from the perspective of an aerospace engineer. Given the need to design for space environments, the chapter examines the effects of non-terrestrial conditions on the various materials used for spacecraft design and on actual humans in space. It presents descriptive data on various historic spacecraft.

Introduction

Aerospace Engineering, as a technical field, was born of the need to design for environments outside of the normal terrestrial conditions. Specialized tools, techniques, design rules and materials are used. The inability to provide repair services on-site in space, with the notable exception of the Hubble Space Telescope repair missions (euphemistically termed refurbishment missions) drives an entire discipline.

Engineered structures existing in space for any period of time will be subject to environmental conditions very different from those on Earth. The effects of vacuum and radiation, for example, on various materials used for spacecraft design depend on the material properties such as density, atomic number, conductivity, and length of exposure and these effects may impact performance. Spacecraft lack the radiation shielding afforded by Earth's atmosphere. The radiation exposure encountered by structures in space causes damage by a variety of mechanisms including charging, electronics degradation, and radiation-induced sublimation. Plasma, ionizing radiation, micro meteor/orbital debris, neutral gases, and the solar and thermal environments in space each have their own effects on materials

M. Ann Garrison Darrin (✉)
Johns Hopkins University Applied Physics Laboratory, Laurel, MD, USA
e-mail: ann.darrin@jhuapl.edu

© Springer International Publishing Switzerland 2015
B.L. O'Leary and P.J. Capelotti (eds.), *Archaeology and Heritage of the Human Movement into Space*, Space and Society, DOI 10.1007/978-3-319-07866-3_2

and structures. These environmental conditions and their effects on engineered structures in space will be explored in this chapter. This is relevant to how the material culture in space will be observed and understood by archaeologists working in the field of space archaeology and heritage.

The Space Environment

The space environment is quite different from that on Earth. It is in many ways harsher and thus spacecraft are designed for the specific environment(s) they will encounter. Radiation effects can be minimized by proper design; for example, critical instruments and spacecraft components can be shielded. Environmental effects can cause failures or contribute to degradation after unrelated spacecraft failures. In general, the spacecraft can interact with the space environment in ways that can cause failure, damage, or affect the integrity of measurements being taken by the spacecraft's instruments. For example, outgassing of materials from spacecraft thermal blankets could leave a fine film on optical components, inhibiting or degrading performance. Sometimes the data disruption can be a way of learning indirectly about spacecraft environment interactions. Remote observations are often the only forensic opportunities when there are anomalies. Much attention has been paid to defining and understanding the space environment, specifically for the purposes of ruggedizing spacecraft design (Bedingfield et al. 1996; Larson and Wertz 1999). Understanding the predicted effect of the environment on the material remains is a major step in following the scientific method for evaluation and analytical purposes of significant remains.

Significant Objects: Terrestrial and Exoatmospheric

For a discussion of the effect of the space environment on materials in space, one can begin with those cultural remains of historical significance that remain terrestrial bound. Often they reflect social or political impacts on the space program, but these artifacts also serve to remind the reviewer of the difficulty of placing materials in space. On December 6, 1957 the United States attempted its first launch of a satellite (*Vanguard TV3*) into orbit, but the launch was a failure. 2 s after leaving the launch pad at Cape Canaveral, this rocket lost thrust and sank back down, rupturing and exploding its fuel tanks, having reached a height of about 4 ft. The Vanguard satellite it was carrying was thrown clear, its transmitters still signaling. It is now on display at the Smithsonian's Air and Space Museum.

Vanguard TV3 was launched in direct response to the surprise launch of *Sputnik 1* on October 4, 1957. *Sputnik's* launch inspired the U.S. to restart the Explorer program, which had been proposed earlier by the Army Ballistic Missile Agency (ABMA); together with the Jet Propulsion Laboratory (JPL) ABMA built

Explorer 1 and launched it on January 31, 1958. Once again, however, the US had been beaten into space by the Soviet Union's second launch, *Sputnik 2*, on November 3, 1957. The televised failure of *Vanguard TV3* on December 6, 1957 only aggravated American dismay over the country's position in the Space Race. Just 3 months later, on March 17, 1958, Vanguard 1 became the second artificial satellite successfully placed in Earth orbit by the United States. It was the first solar-powered satellite; just 152 mm (6 in) in diameter and weighing just 1.4 kg (3 lb.), *Vanguard 1* referred to by -Soviet Premier Nikita Khrushchev as, "the grapefruit satellite" (U.S. Navy 2003; see Fig. 1.1).

Vanguard 1 is the oldest artificial satellite still in space, as Vanguard's predecessors, *Sputnik 1, Sputnik 2,* and *Explorer 1,* have fallen out of orbit. In contrast to *Vanguard TV3,* which is preserved and available for viewing, *Vanguard 1* is preserved in situ, one of the most important relics of the Space Race.

The exploration and exploitation of space reflects both failure and success. To date, 12 men have walked on the moon, and autonomous robotic spacecraft have explored the surface of Mars. Sustained missions to Jupiter and its moons, to Saturn, Titan, Venus and Mercury have been conducted by the US, Europe, and Russia/USSR and the list of countries going to space is growing with China's recent *Chang'e* (lunar lander and rover) and India's *Mangalyaan* (Mars orbiter). Spacecraft have met asteroids, smashed into comets, and orbited the Sun. Two missions have passed the distant outer planets and one of them—*Voyager 2*—is right now crossing a strange boundary that defines the beginning of interstellar space (see Fig. 4.1). *Voyager 2* at a distance of 103.00 AU (1.541 × 1,010 km) from Earth as of 15 November 2013, (Where are the Voyagers, 2013) is one of the most distant human-made objects (along with *Voyager 1, Pioneer 10* and *Pioneer 11). Voyager 2* is part of the Voyager program with its identical sister craft *Voyager 1*; they are in extended mission, tasked with locating and studying the boundaries of the Solar System, including the Kuiper belt, the heliosphere and interstellar space.

Terrestrial Versus Space Preservation

For most Earth-bound archaeological artifacts, features, and sites, the principal form of deterioration arises from the interactions with the atmosphere and humankind that occur unless they are formally preserved and protected. Examples of this deterioration abound for structures on Earth. The Egyptian pyramids, for example, show the effects of millennia of erosion caused primarily by wind-borne particles. In addition, many limestone structures have been harmed by the acidic properties of the air and the resulting low pH of rain. Finally, the degradation of many construction materials arises from oxidation. High levels of tropospheric ozone arising from urban air pollution greatly accelerate this process. Numerous Cold War-era missile silos have been left to degrade through natural processes, and some have even been recycled. In the wetlands near Homestead, Florida there stands a silo, plunging 180 ft straight down into the Earth. Inside that silo stands a rocket, an apparition ten stories high and as wide as a two-car garage. It is the largest solid rocket motor ever built and was intended to take us to the Moon. Stopped by both a technology shift in propellant technology and program cancellations, this behemoth has been sealed underneath concrete highway beams and is at the mercy of the swamp-like environment (Schneider 2010).

In the extraterrestrial realm, material remains are confined to orbiting objects, plus the relatively few pieces of hardware that have successfully landed on the surfaces of the Moon, the inner planets, the moons of planets, and possibly asteroids. The large outer planets (Jupiter, Saturn, Uranus, and Neptune) do not possess clearly defined surfaces, and, in fact, those atmospheres can be a method of disposal for planetary protection purposes, making the likelihood of ever recovering cultural objects that enter their atmosphere minimal.

This same problem appears for orbiting spacecraft that have controlled or uncontrolled—through the Earth's atmosphere. Of great historical significance was the *Mir* space station, owned by the Soviet Union and then by Russia, operated in low Earth orbit from 1986 to 2001 and is of great historical significance. *Mir* was the first modular space station and was assembled in orbit from 1986 to 1996. It had a greater mass than that of any previous spacecraft and held the record for the largest artificial satellite orbiting the Earth until that record was surpassed by the International Space Station after *Mir's* deorbit on 21 March 2001. Near the end of its life, there were plans for private interests to purchase *Mir,* possibly for use as the first orbital television/movie studio, but these plans did not come to fruition as no funding from state or private resources was available (Harland 2004; Dismukes 2010). *Mir's* deorbit was carried out in stages. Reentry into Earth's atmosphere of the 15-year-old space station occurred near Nadi, Fiji. Major destruction of the station began when most of the unburned fragments fell into the South Pacific Ocean (Leamanczyk 2013). A single piece of the Mir's remains was discovered near Boston, MA.

Where Does Space Begin?

There is no firm boundary where space begins. The Kármán line, at an altitude of 100 km (62 mi) above sea level, is conventionally used as the start of outer space for the purpose of space treaties and aerospace records keeping. The Outer Space Treaty, passed by the United Nations in 1967 and ratified by over 100 countries, precludes any claims of national sovereignty and permits all states to explore outer space freely. In 1979, the Moon Treaty made the surfaces of objects such as planets, as well as the orbital space around these bodies, the jurisdiction of the international community.

The baseline temperature, as set by the background radiation left over from the Big Bang, is only 2.7 K; in contrast, temperatures in the coronae of stars can reach over a million Kelvin. Plasma with an extremely low density (less than one hydrogen atom per cubic meter) and high temperature (millions of kelvin) in the space between galaxies accounts for most of the baryonic (ordinary) matter in outer space.

Intergalactic space takes up most of the volume of the Universe, but even galaxies and star systems consist almost entirely of empty space.

Humans began the physical exploration of space during the 20th century with the advent of high-altitude balloon flights, followed by the development of single and multi-stage rocket launchers. Earth orbit was first achieved by Yuri Gagarin in 1961 and unmanned spacecraft have since reached all of the known planets in the Solar System. Outer space represents a challenging environment for human exploration because of the dual hazards of vacuum and radiation. Microgravity has a deleterious effect on human physiology, resulting in muscle atrophy and bone loss.

Space Environment and Interactions with Spacecraft: Environmental Effects on Cultural Material

For the purposes of this section, the space environment will be discussed as it pertains to spacecraft degradation. The relevant environmental factors discussed will be the plasma, ionizing radiation, micrometeoroids and orbital debris, neutral particles, solar, and thermal environments.

The space radiation environments have been modeled and measured for decades (Hastings and Garrett 1996). This work has been documented and summarized of the available models and data on solar protons, heavy ions, and trapped radiation belts, including trapped particle model development over several decades and the Combined Release and Radiation Effects Satellite (CRRES) mission magnetic storm data (Barth 2009).

Plasma

The plasma environment that a spacecraft encounters depends on its orbit and primarily affects the spacecraft through surface charging processes. Spacecraft charging is a process by which a current balance is achieved between currents into and out of the spacecraft. Current in is due to plasma striking the spacecraft surface, and current out depends on the specific materials and their electrical properties. The sources of current out of the spacecraft include secondary electron emission due to electron and ion impacts, backscattered electron emission, and photoemission. Effects on spacecraft include the possibility of arc discharging when two regions of the spacecraft (e.g. sunlight and shaded) charge to different voltage potentials, along with enhanced contamination, shifted spacecraft electrical ground (a problem for measurement integrity), distortion of low-energy particle trajectories, and effects on drag and electromagnetic torque experienced by the spacecraft.

Surface charging due to plasma interactions with the spacecraft is typically minimized by making spacecraft surfaces all electrically conductive. For example, Kapton can be metallized to produce a Faraday cage effect on the surface of a spacecraft. When non-conductive surfaces are required due to other design constraints, the degree of surface charging may be estimated using modeling software.

Ionizing Radiation

Charged particles including protons, electrons, and heavy ions have the potential to cause spacecraft internal charging, degrade electronics, and single-event upsets (SEUs). An SEU can result from a high energy particle impacting and ionizing a sensor element or electronic circuit. Shielding is used to mitigate the effects of ionizing radiation, and its effectiveness can be modeled and measured post-mission. This radiation can also cause dielectric charging and breakdown via arc discharging, degrade optical materials by various processes including color center formation, and have a negative impact on the health of humans in space. Solar arrays are particularly susceptible to radiation effects, which in turn can impact the mission via loss of power generated by damaged solar panels.

Micrometeoroids and Orbital Debris

Macroscopic particles exist near planets due to interplanetary meteoroids and debris. Man-made debris exists near Earth due to decades of presence in the space surrounding Earth and a lack of adequate disposal mechanisms for "space junk." Material from previous satellites, launches, spacecraft, etc. surrounds Earth and can cause physical surface damage when impacting a surface.

These effects can be understood through modeling, and various accounts of damage due to particulate impacts have been recorded in past missions (Sample et al. 2007). Human-made debris—discarded materials or structures—are perhaps what first come to mind when thinking of "space archaeology," as these are the material remains left by humans in outer space.

Neutral Gases

Neutral gases encountered by a spacecraft, predominantly in low-Earth (LEO) and polar Earth orbits (PEO) can cause atmospheric drag, surface erosion, and spacecraft glow, as well as contamination due to neutral gases emitted by spacecraft. The impacting of neutral gas molecules on spacecraft in LEO transfers momentum and energy to the spacecraft resulting in drag, which can impact orbital precision.

Contamination due to neutral gases can originate from outgassed material from the spacecraft itself or in the form of propellant molecules from solid rocket motor propulsion systems condensing on the spacecraft. This material can degrade performance of the spacecraft or its instruments. Great care is taken in spacecraft design to use materials which are "flight qualified"—that is, whose outgassing properties under stringent limitations are known. Materials can also be thermally pre-treated on earth to minimize outgassing in space.

Neutral gases also include atomic oxygen which degrades spacecraft surfaces via erosion and recession. This effect was seen on space shuttle flights in the 1980s (Tribble 2003). Oxygen is a strong oxidizing agent and is a significant component of the LEO atmosphere. Erosion is enhanced by ultraviolet (UV) exposure, and is most significant with many organic materials. Spacecraft glow is also a neutral gas effect related to contamination and can spectrally affect optical sensor system measurements.

Solar Environment

The sun's electromagnetic flux and emitted charged particles are its main contributions to space environmental effects. The sun emits high energy protons and electrons as well as lower energy plasma known as the solar wind. Solar activity variations result in solar flares and geomagnetic storms. Missions are often planned so as to avoid major solar events which could compromise the spacecraft, an astronaut's health walking in space, ability to acquire data, or data integrity. The sun's photons also cause environmental effects in the form of photoemission and electron generation via processes such as Compton scattering and pair production.

Thermal Environment

The thermal environment and its effects on aging are exacerbated in the temperature extremes of space. Briefly, space instruments and electronics have operational temperature ranges close to ambient Earth temperatures. Care must be taken in space to keep these electronics within their operational temperature ranges, which is accomplished both by spacecraft design and also via materials such as thermal blankets. These materials are susceptible to aging and degradation in space. When referring to the exoatmosphere the Kelvin (K) temperature scale is primarily used. The Kelvin scale is designed so that 0 K are defined as absolute zero, the hypothetical temperature at which all molecular movement stops. The size of one unit is the same as the size of 1 °C. The background of space at 2.7 K is equivalent to −454.81 °F or −270.45 °C.

Natural Versus Induced Environments

The environments covered thus far are natural, that is, defined as the space environment as it occurs independently of a spacecraft's presence. Orbital debris, while manufactured by humans at some point, may be present as part of the natural environment to a new mission. Spacecraft operation may also induce a local environment not representative of that which the craft passes through.

Many examples of this exist throughout the history of our space exploration, and range from the somewhat mundane "water" dumps from the space shuttle to plasma cloud seeding experiments designed to generate plasma for study. Retrorocket fire from the space shuttle approaching the space station is another source of induced environment, as well as any local thermal perturbations due to the presence of the spacecraft. The classic example of an induced environment is contamination. Contamination can be molecular, in the example of outgassed molecules recondensing onto windows or solar cells, or particulate, in the example of particles released into the local environment by vibrations of the spacecraft or its parts.

Combined Effects

All of the space environmental effects described here may coexist together during a given mission. Many act in combination, leaving the spacecraft more vulnerable to problems. For example, the degree of outgassing due to the vacuum (microgravity) environment is typically exacerbated by elevated temperature.

However, other combinations of effects may mitigate overall environmental effects. For example, increased temperature typically renders insulating materials more conductive, thereby possibly providing a mitigating effect on differential spacecraft charging.

Combinations of these effects are obviously as specific to the mission as the particular environments encountered and materials used. However, it is necessary to keep in mind that effects may combine, producing unanticipated effects to the spacecraft or mission data integrity.

The combined space environmental effects on potential Apollo spacecrafts' window materials have been studied in detail. The study identified that the primary environmental factors contributing to the degradation of the optical windows were ultraviolet (UV) and extreme ultraviolet (EUV) photons, solar wind protons, any solar flare protons and electrons, and protons and electrons trapped in the Van Allen Belts (Pigg and Weiss 1973). Optical transmissivity was decreased through many of the potential window materials to a degree that depended on the duration of exposure. Often, terrestrial studies such as this one can be useful to help spacecraft engineers determine which materials to use for a given mission.

Effects of Space Environment on Humans

Combined space environmental effects are particularly relevant to space travelers such as astronauts or tourists. Just as spacecraft materials are exposed to the environment, so is the human body in outer space, particularly on a spacewalk. Periods of weightlessness (due to the microgravity environment) are known to reduce the body's bone mass, and also may depress the immune system and lead to changes usually associated with aging. In addition to the microgravity environment, the radiation environment presents many potential hazards. Ionizing radiation can damage cell DNA, potentially preventing cells from being able to self-repair, or may cause mutations. Radiation can cause acute sickness, depending on the dose, and solar flare incidents would be particularly hazardous to the unprotected astronaut. Humans are generally protected from many types of radiation exposure by the earth's environment, and from the shielding capability of structures such as a spacecraft or space station.

Planetary: Surfaces and Atmospheres

Earth: The Earth's atmosphere is one of the reasons why life can survive on our planet. The thick atmosphere is composed of common elements such as nitrogen (78 %), oxygen (21 %), and argon (1 %). The rest is comprised of very small amounts of neon, helium, methane, carbon dioxide, krypton, hydrogen, xenon, ozone, nitrous oxide, carbon monoxide, ammonia, and iodine. There is also water vapor in the lower part of Earth's atmosphere.

Earth's Moon: The thin atmosphere of our Moon means that surface artifacts are exposed to extreme UV radiation, extreme temperatures from the monthly cycles of exposure to solar heating and then deep space and finally to the solar wind. There are reactive chemical species in the atmosphere because exposure to extreme UV initiates photochemistry that forms radicals, but the number density of these reactants is too low for them to be factors in the deterioration of archeological objects. The debris field from the Apollo 11 mission has been exposed to severe temperature cycles as well as to extreme UV. So any organic remains (such as refuse from Neil Armstrong's bodily functions—a popular internet topic) left on the surface would be, by now, simply long-sterilized dry dust.

Mercury: Mercury's atmosphere is very thin and highly variable. Essentially, it is a surface-bound exosphere containing hydrogen, helium, oxygen, sodium, calcium, potassium and water vapor, with a combined pressure level of about 10–14 bar (1 nPa). The exospheric species originate either from the solar wind or from the planetary crust. Solar pressure pushes the atmospheric gases away from the Sun, creating a comet-like tail behind the planet.

Venus: Venus has an extremely thick atmosphere blanketing the planet's surface. Because the thick atmosphere obscures the planet's surface, scientists and writers alike speculated that the planet was covered with lush forests. Actually, the atmosphere hides a barren, burning planet. The thick clouds of Venus are composed mostly of toxic carbon dioxide, so that solar heating of Venus is not effectively reradiated into space and the surface temperature exceeds 755 K (900 °F). The surface pressure is on the order of 90–100 bars, or about 100 times the pressure at Earth's sea level. With these pressures material remains of spacecraft such as the Soviet *Venera* missions would be crushed in a short time (minutes to several hours). The scale height is about 16 km. Although there is little free oxygen on Venus, there are substantial amounts of carbon monoxide which is somewhat reactive as well as sulfuric acid in cloud droplets which would attack most metal objects during a descent to the surface.

Mars: Like Mercury, Mars has a thin atmosphere. The atmosphere of Mars is poisonous to humans because it is mostly (95 %) composed of carbon dioxide. The rest of it is nitrogen (3 %), argon (1.6 %), and very small amounts of water, methane, and oxygen. Mars does experience very significant dust storms of long duration and with high wind velocities. Particle sizes are estimated to be small—on the order of 1.5 microns in diameter. The wind velocities, driven by the summer-winter freezing and thawing of significant portions of the atmosphere can reach 400 km/h (250 mph). However, the surface pressure on Mars is only about 7 mB and so the amount of abrasive dust that could be lofted by such a tenuous atmosphere is probably small, meaning that physical erosion should be slow. On the other hand, the thin atmosphere does mean that the surface of Mars is exposed to intense solar UV radiation and considerable temperature extremes. These latter two problems are more likely to cause stress to a human-made object such as rovers on Mars.

Jupiter: As a gas giant, Jupiter does not actually have a surface or a well-defined end to its atmosphere. Jupiter is composed of elements—hydrogen (90 %) and helium (10 %)—that are typically considered components of an atmosphere. There are also some trace amounts of other molecules, such as methane, ammonia, water, and hydrogen sulfide. Scientists have defined the lower end of Jupiter's atmosphere as the point where the pressure is 1 bar.

Saturn: Like Jupiter, Saturn is composed of mostly helium and hydrogen and does not have a defined surface; the atmosphere also contains traces of methane, ammonia, water ice, and other compounds. Saturn's upper atmosphere is mostly ammonia crystals while the lower one is either water or ammonium hydrosulfide.

Some of the different elements in the atmosphere combine to form what we call smog in our atmosphere.

Uranus: Uranus appears as a blue-green orb in space. The color comes from the methane in Uranus' atmosphere; the substance absorbs red wavelengths, so they do not reflect back into space. Like other gas giants, Uranus is mostly composed of hydrogen and helium; however, unlike Jupiter and Saturn, Uranus has a higher concentration of what are known as "ices"—mixtures of compounds that include ammonia, ammonium, methane, and water.

Neptune: Neptune is also a gas giant, but it has a higher proportion of ices than any other planet in our Solar System. Like Uranus, Neptune appears blue as a result of the methane in its atmosphere. Like the other gas giants, Neptune's atmosphere has different storms. At up to 2,400 km/h, Neptune's winds are the fastest of any of the planets.

Fig. 2.1 Mariner 2 sped past Venus with radar instruments that confirmed that the cool cloud cover of Venus concealed a surface hot enough to melt lead. The flyby was humanity's first visit to another planet and revealed a world caught up in a runaway greenhouse effect (NASA)

Case Studies: The *Mariner 2* Sixty Years Later and Legacy of *Lunokhod 1* and GRAIL

Mariner 2

On 14 December 1962, a NASA spacecraft called *Mariner 2* sped past Venus with radar instruments that confirmed that the cool cloud cover of Venus concealed a surface hot enough to melt lead. The flyby was humanity's first visit to another planet and revealed a world caught up in a runaway greenhouse effect (Fig. 2.1).

The two-stage Atlas-Agena rocket carrying *Mariner 1* had veered off-course during its launch on July 22, 1962 due to a defective signal from the Atlas and a bug in the program equations of the ground-based guiding computer, and subsequently the spacecraft was destroyed by the Range Safety Officer. A month later, the identical *Mariner 2* spacecraft was launched successfully on August 27, 1962, sending it on a 3½-month flight to Venus. On the way, it measured the solar wind, a constant stream of charged particles flowing outwards from the Sun, confirming the measurements by *Luna 1* in 1959.

It also measured interplanetary dust, which turned out to be scarcer than predicted. In addition, *Mariner 2* detected high-energy charged particles coming from the Sun, including several brief solar flares, as well as cosmic rays from outside the Solar System. As it flew by Venus on December 14, 1962, *Mariner 2* scanned the planet with its pair of radiometers, revealing that Venus has cool clouds and an extremely hot surface.

Mariner 2 did not have the catastrophic failure of *Mariner 1*, but it did see numerous anomalies, primarily due to the space environment. On September 8, 1962, the spacecraft automatically turned on its gyros and automatically turned off the cruise science experiments. The exact cause is unknown since attitude sensors went back to normal before telemetry measurements could be sampled, but possibly an Earth-sensor malfunction or a collision with a small unidentified object temporarily caused the spacecraft to lose Sun lock.

On September 29, 1962, all sensors went back to normal before it could be determined which axis had lost lock. Shortly thereafter, on October 31, the output from one solar panel (with a solar sail attached) deteriorated abruptly. The diagnosis was a partial short circuit in the panel, and the cruise science instruments were turned off as a precaution. A week later the panel resumed normal function and cruise science instruments were turned back on. The panel permanently failed on November 15, but by then *Mariner 2* was close enough to the Sun that one panel could supply adequate power; thus the cruise science experiments were left active. Despite issues stemming from multiple causes, from design flaws to space dust, *Mariner 2* was still able to complete its mission. Since all evaluations were done remotely and in general were not verifiable, the true causes will never be validated. The spacecraft's last transmission was January 3, 1963 and it is now defunct in heliocentric orbit.

Lunokhod 1

Lunokhod was a series of Soviet robotic lunar rovers designed to land on the Moon between 1969 and 1977. *Lunokhod 1* landed in 1970, followed by *Lunokhod 2* in 1973 at the height of the Cold War. The successful missions were in operation concurrently with the Zond and Luna series of Moon flyby, orbiter and landing missions. The Lunokhods were primarily designed to support the planned Soviet manned moon missions and to be used as automatic remote-controlled robots to explore the surface and return pictures.

In 2010, nearly 40 years after the 1971 loss of signal from *Lunokhod 1*, the NASA Lunar Reconnaissance Orbiter photographed its tracks and final location, and researchers, using a telescopic pulsed-laser rangefinder, detected the robot's retroreflector (Bleicher 2010). For nearly 40 years the exact location of *Lunokhod 1* was only known within a few kilometers. Attempts to find the rover by shooting a laser against the rover's 14 silver coated corner cube retroreflectors from Earth were difficult, due to the effects of Earth's atmosphere and perhaps complicated by a coating of lunar dust on the rover.

Specialists at the Apache Point Observatory Lunar Laser-ranging Operation (APOLLO) in southern New Mexico used the LRO images to first pinpoint the

Fig. 2.2 Lunokhod 1 model on the left contrasted to STR-1 designed for work at the Chernobyl site. In order to survive the lunar surface, the rovers such as Lunokhod were designed for the radiation environment of space—a design that later paid off on Earth. After Chernobyl, the Lunokhod designers were called back to work to create rovers that could operate in the post-disaster environment. In just 2 weeks, rovers were made which used nuclear decay heat sources for internal rack climate control and had electronic systems that were already hardened to resist radiation (author composite)

locale of *Lunokhod 1* with sufficient positional accuracy to permit laser range measurements. Surprisingly, the APOLLO researchers reported that the craft's retroreflector was returning much more light than other reflectors on the moon. *Lunokhod 1* has returned to providing scientific data over 40 years later (Fig. 2.2).

In order to survive the lunar surface, the rovers were designed for the radiation environment of space—a design that later paid off on Earth. After Chernobyl, the Lunokhod designers were called back to work to create rovers that could operate in the post-disaster environment. In just 2 weeks, rovers were made which used nuclear decay heat sources for internal rack climate control and had electronic systems that were already hardened to resist radiation (Museum of Robotic Equipment). The Lunokhod model provided a capability that traditional Earth-based robots could not, allowing operation in the high radiation environment (Fig. 2.3).

GRAIL

The GRAIL mission placed two spacecraft into the same orbit around the Moon in 2011. As they flew over areas of greater and lesser gravity, caused both by visible features such as mountains and craters and by masses hidden beneath the lunar surface, they moved slightly toward and away from each other. An instrument aboard each spacecraft measured the changes in their relative velocity very precisely, and scientists translated this information into a high-resolution map of the Moon's gravitational field.

At the end of the science phase and a mission extension, the spacecraft was powered down and decommissioned over a five-day period. The two spacecraft impacted the lunar surface on December 17, 2012. Both spacecraft impacted an unnamed lunar mountain between Philolaus and Mouchez at 75°37′N 26°38′W

Fig. 2.3 The Soviet Union's Lunokhod 1 lunar rover in its final parking spot. Part of the Luna 17 mission, this first successful rover operated for 11 lunar days, the equivalent of 322 Earth days. It traveled more than 10 km across the lunar surface (NASA)

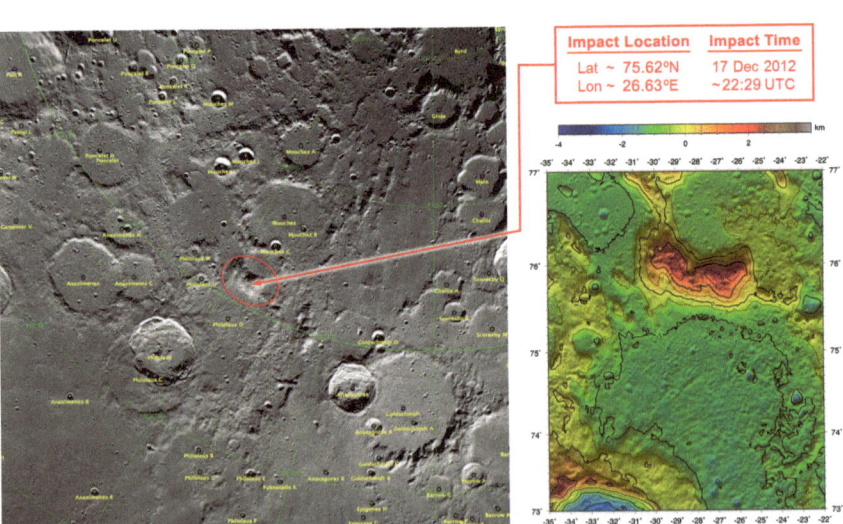

Fig. 2.4 These maps of Earth's moon highlight the region where the twin spacecraft of NASA's Gravity Recovery and Interior Laboratory (GRAIL) mission impacted the moon marking the end of its successful endeavor to map the moon's gravity. The two washing-machine-sized spacecraft, named Ebb and Flow, impacted at a then unnamed mountain near the moon's North Pole (NASA/GSFC)

and 75.62°N 26.63°W. Ebb, the lead spacecraft in formation, impacted first, and Flow impacted moments later. At impact, each spacecraft was traveling at 3,760 miles/h (6,050 km/h). Before impact, the main engines aboard each spacecraft were fired in order to deplete the remaining fuel. The crash site has been named after GRAIL collaborator and first American woman in space, Sally Ride. In effect, this is another recent archaeological sites (Fig. 2.4).

The Grail mission ended Dec 17, 2012 with a bang rather than a whimper. But Mariner 2, cold and silent for five decades, still orbits the sun: a reminder of bygone marvels, and maybe also a reproach to the future.

Conclusions

The archaeological study of the material remains in space allows us to stretch our minds past the terrestrial cultural landscape to the concept of a cultural "spac-escape." This chapter serves as a cursory introduction to environments which are exoatmospheric and removed from Earth's protective blanket (our own magnetosphere).

In contemplating space archeology, what once was seen as inhospitable environment to material remains on our planet actually allows us to appreciate how protected we are on our blue marble. The physically and chemically erosive elements such as air, erosion and caustic pollutants are relatively tame compared to the cosmic radiation and dramatic temperature extremes of space.

The famous Earthrise image, taken by astronaut William Anders in 1968 during the Apollo 8 mission, was perhaps the most influential environmental photo ever and has taught us humility as we understand our very precious space in our solar system. The fact that humankind has stepped outside the protective blanket of the Earth is, in itself, of such great significance. As J. G. Ballard wrote in an article titled "Back to the Heady Future," (Daily Telegraph, 17 April 1993), "Neil Armstrong may well be the only human being of our time to be remembered 50,000 years from now."

The study of space archaeology, although still in its infancy, requires us to contemplate and plan for where we will be in space and its future historical and cultural significance. These criteria must be considered as an essential part of activities related to planning for both space exploration and exploitation.

References

Ballard, J. G. (1993). Back to the heady future, daily telegraph, April 17, 1993.
Barth, J. L. (2009). Space, atmospheric, and terrestrial radiation environments. In A. Darrin & B. O'Leary (Eds.), *Handbook of space engineering, archaeology and heritage* (pp. 529–560). Boca Raton: CRC Taylor and Francis Press.
Bedingfield, K. L., Leach, R. D., & Alexander, M. B. (1996). *Spacecraft system failures and anomalies attributed to the natural space environment.* Washington, DC: NASA RP-1390.

Bleicher, A. (2010). Forgotten soviet moon rover beams light back to Earth, IEEE Spectrum. Posted 18 Aug 2010. http://spectrum.ieee.org/aerospace/robotic-exploration/forgotten-soviet-moon-rover-beams-light-back-to-earth

Dismukes, K. (2004). Accessed 12/12/2013. http://spaceflight1.nasa.gov/history/shuttle- mir/history/history.htm

Harland, D. (2004). *The story of space station mir*. New York: Springer.

Hastings, D. E., & Garrett, H. B. (1996). *Spacecraft environment interactions*. New York: Cambridge University Press.

Larson, W. J. & Wertz, J. R. (1999). *Space mission analysis and design* (3rd ed. Vol 8). Portland, Oregon: Space Technology Library.

Leamanczyk, L. (2013). "Rock found in amesbury backyard came from space station," 14 June 2013. Accessed Jun 16 2013. http://boston.cbslocal.com/2013/06/14/rock-found-in-amesbury-backyard-came-from- space-station/

Pigg, O. E., & Weiss, S. P. (1973). *NASA technical note*, NASA TN D-7439.

Sample, J. L., Donegan, M., Wolf, T., Drewry, D., & Mehoke, D. (2007). Charging of ceramic coatings in space. In *48th AIAA/ASME/ASCE/AHS/ASC Structures, Structural Dynamics, and Materials Conference Proceedings*. Waikiki, Hawaii.

Schneider, D. (2010). Hidden in the glades, a giant relic of the U.S. quest for space. Accessed Nov 15 2013. http://www.keysnet.com/2010/06/10/227720/hidden-in-the-glades-a-giant-relic.html

Tribble, A. C. (2003). *The space environment*. Princeton, NJ: Princeton University Press.

U.S. Navy. (2003). Vanguard Satellite Marks 45 Years in Space, U.S. Navy Press Release, 10 March 2003. Accessed Jan 3 2014. http://www.nrl.navy.mil/media/news-releases/2003/vanguard-satellite-marks-45-years- in-space

Chapter 3
Robot Avatars: The Material Culture of Human Activity in Earth Orbit

Alice Gorman

Abstract This chapter discusses orbital debris from a cultural heritage perspective. It examines the cultural material related to space exploration with a specific focus on "space junk" and the increasing amount of material remains including thousands of satellites, rocket bodies, parts and piece of spacefaring objects. The author argues that the materials and design reflect social and political interactions with space as well as humanity's adaptation to a new environment. The study of space heritage can add to the history of space exploration and contemporary life on Earth.

Introduction

Since the launch of *Sputnik 1* in 1957, more than 4,900 rockets have delivered payloads into Earth orbit that now perform a critical role of the fabric of modern life (ESA 2013). There are few nations that do not depend on satellite-delivered navigation, telecommunications and Earth observation data, across both civil and military domains. At the individual level, the terrestrial infrastructures which provide mobile phones, navigation and the internet are facilitated by access to satellites. For the contemporary entity, life without access to Position, Navigation and Timing (PNT) and Global Navigation Satellite Systems (GNSS) is rapidly becoming unthinkable.

This chapter discusses the broad research questions that can be addressed by considering the thousands of satellites, rocket bodies, and pieces of junk currently in Earth orbit as a cultural landscape or assemblage. Their materials and design reflect the nature of our social and political interactions with space, as well as adaptations to a new environment. Factors that contribute to the character of this material record include microgravity, extreme temperature and radiation conditions, national

A. Gorman (✉)
Flinders University, Adelaide, Australia
e-mail: alice.gorman@flinders.edu.au

© Springer International Publishing Switzerland 2015
B.L. O'Leary and P.J. Capelotti (eds.), *Archaeology and Heritage of the Human Movement into Space*, Space and Society, DOI 10.1007/978-3-319-07866-3_3

political and scientific agendas, and technological styles through time and across terrestrial cultures. Space junk has a story to tell about contemporary life on Earth.

Although humans have made forays into this challenging environment on orbital and lunar missions, it remains essentially a robotic frontier. Unlike terrestrial artifacts, satellites in orbit are barely visible from Earth and are not designed to interact with human bodies. They are remote, separated from us in space and sometimes even in time (for example, the *Voyager 1* and *2* deep space probes) and remotely controlled, yet highly autonomous, as they can frequently operate without direct intervention for long periods (Goodrich and Schultz 2007: 239).

Non-operational spacecraft and fragments of spacecraft are called 'junk' or debris. One commonly used definition of space junk is a piece of hardware that does not currently, or in the foreseeable future, serve a useful purpose (Crowther 1994: 128; Mehrholz et al. 2002; NASA 2012). As with other resources such as minerals, this is a value judgment based on the present lack of capability to salvage, recycle, re-purpose, or otherwise find a use for this material which has been lofted into orbit at such great expense (approximately 20,000 USD per kg; NASA 2008). Taking a longer-term view, space junk may represent the beginnings of a technological trajectory that will transform how human cultures relate to time and space.

Space Junk in Earth Orbit

The number of spacecraft and artifacts in orbit is far greater than the number of launches. Typically, rockets designed for this purpose are multistage, and release the payload from an upper stage, which often remains in orbit. Despite the frequent re-entry of spent rocket bodies due to orbital decay and atmospheric drag, Low Earth Orbit (LEO) is littered with Agenas, Deltas, Ables, Scouts, and many others. Rockets frequently release multiple payloads, and the release involves the separation of protective fairings and other objects such as bolts and lens caps (mission-related debris).

Once in orbit, spacecraft are exposed to a high radiation environment of temperature extremes. Over time, the effect of atomic oxygen, cosmic rays, plasma arcing, and bombardment with "natural" objects and other space junk can cause the spacecraft materials to decay and degrade. But the primary source of fragmentation debris derives from the explosion of upper rocket stages with residual fuel (NASA 2012). Over 250 on-orbit fragmentation events have occurred since 1961 (Liou and Anz-Meader 2010; ESA 2013). This includes the infamous 2007 Chinese anti-satellite missile test, which destroyed the defunct polar-orbiting *Fengyun 1C* satellite. The *Fengyun 1C* incident was the most catastrophic ever to occur in LEO; this single event created 2,377 pieces of tracked debris, which make a significant contribution to the long term environment (Pardini and Anselmo 2009, 2011). In addition to the catalogued fragments, it is estimated that the event generated over 150,000 fragments less than 1 cm.

These taphonomic processes have created more than 21,000 objects that are larger than 10 cm (a general limit of tracking), according to NASA's Orbital Debris Office. Of these catalogued objects, 51 % are non-operational whole spacecraft and 43 % are debris from spacecraft fragmentation. A mere 6 % are functioning satellites. According to the definition of non-useful, 94 % of all tracked items in Earth orbit are technically junk. But this is only the tip of the iceberg. Below the limits of tracking, the estimated population of particles between 1 and 10 cm in diameter is approximately 500,000. The number of particles smaller than 1 cm exceeds 100 million (Mehrholz et al. 2002).

Operational spacecraft and space junk form an archaeological record that is radically different to that on Earth, where sites and artifacts both on the surface and beneath it have their position uniquely described with regard to height above sea level, grid coordinates or latitude/longitude, or XYZ coordinates in archaeological excavations (Gorman 2009a). It is not possible to uniquely describe the position of a piece of space junk as it is in constant motion, travelling at an average speed of 10 km per second (Belk et al. 1997; NASA 2012), and its orbit evolves through time. Orbits must be actively tracked and predicted in order to know location. Location in orbit is described by classical orbital equations with six Keplerian elements, and affected by numerous factors such as electromagnetic environment, solar heating and expansion of the atmosphere, attitude, altitude and various others.

Despite this constant movement, however, space junk is far from randomly distributed throughout near-Earth space. There are particular orbital regimes which have higher densities, because of (1) location of launch site, (2) function of satellite, e.g. Earth observation, telecommunications, PNT, GNSS (3) avoidance of the high radiation environment of the Van Allen belts and (4) location of terrestrial end users, among many other factors. Orbits can be described by both altitude—the height of perigee and apogee above the surface of the Earth—and inclination to the equator. Commonly used orbits are Low Earth, Medium Earth, Geosynchronous, Geostationary (GEO), Sun-synchronous, Molniya, polar, and the disposal or 'graveyard' orbit about 400 km above GEO.

Objects sometimes also get 'discarded' in transfer orbits, such as the Hohman, which is used to make the transition between low and high Earth orbits. Left to themselves, satellites would roam like wayward stars across the dynamic gravitational landscape; 'station-keeping' maneuvers (using the satellite's engines and verniers) are required to keep the functioning spacecraft in its assigned orbit. The spatial patterning of space junk is a result of original mission requirements, post-mission orbital decay, and space environment factors. At the most basic level, this creates relationships between spacecraft that have diverse origins.

The Whole Is More Than the Sum of Its Parts

It is easy to make a case that individual spacecraft have historic, aesthetic, scientific, spiritual and social cultural significance as defined by Australia ICOMOS' internationally recognised heritage guidelines, the Burra Charter

(Australia ICOMOS 2013). Satellites such as *Vanguard 1*, *Astérix 1*, *Telstar 1*, *Syncom 3* and *Australis Oscar V*, which are still in Earth orbit, have played roles in global space exploration by demonstrating new technology or capabilities, and in promoting nationalist space agendas, particularly in the Cold War (e.g. Green and Lomask 1970; Gorman 2005b; Gorman and O'Leary 2007). *Australis Oscar V* represents a lesser-known aspect of space development in this period, the involvement of amateurs (Gorman 2009c; see also McCray 2008).

For each of these we could ask particular research questions: for example, what were the design influences on the spherical *Vanguard 1* (USA, still in orbit), and *Sputnik 1* (USSR, re-entered), as both prototypes were the first of their kind? What could the comparison of the physical bodies of the satellites reveal, that the extensive documentary record cannot, about both the Cold War rivalry and the global cooperative circulation of science in the International Geophysical Year? Even the names of these spacecraft reflect ideological perceptions of space: 'vanguard' is military in origin, indicating the leading position in an advancing army, while 'sputnik' is commonly translated as 'travelling companion'. The irony is that such an investigation must be carried out on models of the spacecraft that exist in various museums, as neither original is available. For the present, when it is not possible to make archaeological field trips to Earth orbit, it barely matters whether the actual spacecraft exists for such a study to be carried out.

Taking each spacecraft as an individual unrelated to the others reduces the orbital population to an aggregate. The aggregate approach has implications for heritage management—for example, a historic spacecraft could be removed from orbit for conservation and museum display without impacting its cultural values or those of other spacecraft (e.g. Schiffer 2013: 173). The cultural significance of orbital debris as an entity is perhaps less obvious. In previous work, I have argued that the totality of human material in orbit is an organically evolved cultural landscape, or spacescape, as defined by the World Heritage Convention, and as such may have a cultural significance distinct from the cultural significance of the individual items which constitute it (Gorman 2005a, 2009b).

In terms of conceptualizing orbital debris, an advantage of the cultural landscape approach is that it moves away from the instrumental view of space as an 'empty' vacuum into which human materials are inserted, foregrounding the otherwise largely invisible background. Orbital debris becomes *part* of near-Earth space rather than time-travelling objects merely passing through without altering a blade of metaphorical grass (Bradbury 1952). The landscape approach treats all objects as equal rather than privileging the 'old', as a surface rather than a stratigraphy.

Archaeologists could also look at orbital debris as an assemblage. In strict archaeological terms, an assemblage is a collection of artifacts from the same context, whether a surface area or an excavation unit. Their association is created by the frame or the analytic unit, and interpreting the relationship between the objects in the assemblage is a matter of taphonomy, middle-range theory and behavioural, evolutionary or social theory. Such assemblages are frequently palimpsests.

An alternative concept of assemblage has been gaining currency in social theory and archaeology, derived from Deleuze and Guattari (e.g. 1987), and further developed by De Landa (2002, 2006); It is important to note that assemblage does not derive from the French word *assemblage*, but is a translation of *agencement* (Philips 2006). This sort of assemblage has emergent properties that are more than the sum of its parts, arising from the interactions between its components; like the landscape, it can include both human and non-human (De Landa 2006: 3–4).

My characterization of the assemblage here is no more than a nod to Deleuze and Guattari's more recursive and metaphorical concept than an actual application, but no less useful for that. The assemblage is constructed through specific histori-cal processes. It bears some relationship to the archaeological assemblage, but also has features of the cybernetic feedback system (e.g. Flannery 1968; Salmon 1978), and non-linear systems in general (e.g. Prigogine and Stengers 1984). Venn (2006) notes the debt of Deleuzian assemblage theory to topology, cybernetics, non-linear turbulence and chaos. This is particularly apt, of course, as the orbital population is the exemplar of the non-linear n-body problem, where n is any number greater than three. The assemblage "is thus a resource with which to address in analy-sis and writing the modernist problem of the heterogenous within the ephemeral, while preserving some concept of the structural" (Marcus and Saka 2006: 102).

Both landscape and assemblage allow the incorporation of the less-than-whole fragments and smaller pieces into the system; our analysis is not confined to whole or nearly whole spacecraft with structural integrity. At the same time, as opposed to the aggregate approach, there is no necessity to assess the cultural significance of every tiny piece (Gorman 2005b). Mission-active spacecraft are as much a part of the assemblage as the defunct and junk (but see Capelotti 2010 for a distinction between archaeological and systemic contexts). The relationships and interactions between objects in orbit are the key to understanding this assemblage as an archae-ological entity.

Sources of Evidence

The main problem in studying orbital debris from an archaeological perspective is the impossibility of direct field experience. The distances, spaces and speeds are vastly greater than terrestrial ones, and at this point in time propulsion tech-nology and delta-v budgets do not allow for the luxury of this kind of mission. Even remote sensing of Earth orbit from the surface of the Earth can only view a swath, so it is impossible to 'see' the entire landscape at any one time. There are, however, other means of perceiving the orbital spacescape. Direct methods include radar, optical and laser tracking, the study of returned spacecraft surfaces, and 'beam-park' sampling. Indirect means include the historical data, simulations and modelling.

A number of space agencies and organizations (e.g. European Space Agency [ESA], Russia, NASA, USSTRATCOM), have orbital debris tracking networks or antennas, and maintain extensive data catalogues. What is tracked depends on size, altitude and cross-sectional area. The higher the orbit, the larger the object needs to be visible. In general, only objects above 5 cm, usually 10 cm, can be tracked, so the catalogues are always a small sample.

There are publicly accessible versions of these catalogues. CelesTrak is run by the Center for Space Standards & Innovation (CSSI), an astrodynamical research division of Analytic Graphics Inc (AGI). Another widely used online catalogue is Heavens Above, maintained by Chris Peat. These databases base their computations of position on Two-Line Elements (TLEs), which encode the more complex six element Keplerian equations. Orbital positions can be calculated from the TLEs using a set of algorithms. However, data derived from TLEs older than 30 days can become unreliable. Other information in the databases includes spacecraft name, date of launch, launching state, and whether it is a whole spacecraft or a debris fragment.

Returned spacecraft surfaces are an important source of information about debris too small to be tracked. These can be used to measure the flux of debris particles, including density, size, speed and angle of collision. Space shuttle windows showed evidence of constant bombardment by tiny particles and these collisions were analysed to gauge the nature of the debris flux in LEO (Hyde 2000a, b).

The Long Duration Exposure Facility (LDEF) was a bus-sized experimental spacecraft with panels of different materials and coatings commonly employed in spacecraft manufacture. Deployed in 1984 by the space shuttle, it stayed in LEO for 5.7 years before retrieval, providing a wealth of data about the debris flux as well as impacts of space environment on materials (e.g. Levine 1991).

Another tool in the characterization of the small-size debris population is 'beam-park' experiment. A radar beam is maintained in a fixed direction with respect to the Earth and all objects that pass through the beam are registered. Depending on the antenna, a 'snapshot' taken over 24 h provides a sample of objects as small as 2 mm, up to an altitude of about 2,000 km. In some cases, the source of previously uncatalogued objects can be identified. The beam-park data is also used to validate orbital debris models (Mehrholz et al. 2002). Both the USA and Europe gather beam-park data.

Models and simulations, both statistical and predictive, of the orbital debris environment are made using tracking and returned surface data. One example is ESA's Meteoroid and Space Debris Terrestrial Environment Reference (MASTER), which describes the spatial distribution and particulate flux as a function of size and location in space, and AGI's Systems Tool Kit (STK, formerly the Satellite Tool Kit).

These mathematical models have to be validated with measurement data such as beam-park. There is a multitude of ways of representing the orbital catalogue data. Gabbard diagrams, for example, plot apogee and perigee height against period (thus each object appears twice). Figure 3.1 shows a Gabbard diagram of almost 300 pieces of debris from the disintegration of the third stage of a Chinese Long March 4 rocket in 2000.

Another form of data visualization shows debris in relation to the Earth. Figure 3.2, from the NASA Orbital debris office, shows the evolution over six months of 2,000 catalogued debris objects from the *Fengyun 1C* break-up. Essentially, over time, they have enclosed the globe.

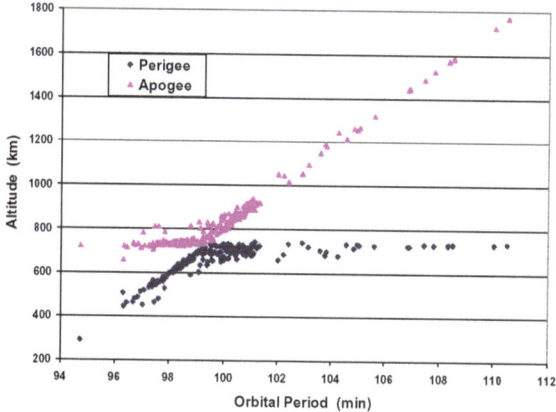

Fig. 3.1 The Gabbard diagram is a way of visualizing the dispersal of debris after a break-up event. For each piece, its apogee (the part of its orbit that is furthest from earth) appears as a pink triangle and its perigee (the part of its orbit that is closest to earth) appears as a blue diamond. While most fragments of the Long March 4 rocket are between 400 and 1,000 km above the earth, some have developed more eccentric orbits with apogees around 1,800 km (NASA)

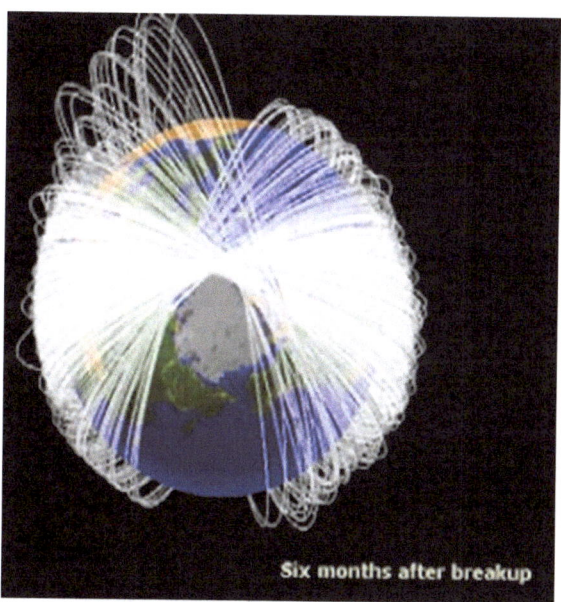

Fig. 3.2 All debris objects start off more or less in the same orbit as the parent object, but diverge over time. In this visualization, each white line represents the orbit of a debris fragment from Fengyun 1-C. Some develop increasingly eccentric orbits, which can also be seen in the Gabbard diagram above (NASA)

Indirect data also includes the documentary, archival and oral history evidence for how spacecraft were manufactured, the materials they were made from, their purpose and dimensions (e.g. fabric, function and size) and their individual histories. This information can be used statistically as well as for researching individual spacecraft. Quirks and individual variations, incidents and events that may have impacted on the final form of the satellite are often recorded formally or informally.

Between direct and indirect sources, there is an abundance of information with which to explore the orbital assemblage and its meanings. In the section below, I use a combination of both: optical data from a satellite tracking camera, together with background research, to demonstrate how we might begin to understand spatial relationships in the terrestrial-orbital system.

Looking at GEO from Earth

The Geostationary orbit, 35,786 km above the Earth, is a critical one for terrestrial telecommunications, as a constellation of only three satellites at this altitude can provide coverage of the entire Earth. Far from the reach of atmospheric drag, spacecraft in this region are likely to remain in orbit for potentially thousands of years. Use of the GEO ring commenced with the successful injection of *Syncom 3* in 1963, and by 2004, there were 1,000 satellites (Jehn et al. 2005: 1326). In 2013, there were 416 active satellites (Johnston 2013).

Figure 3.3 is a photograph taken on November 18, 2012 by Dr Marco Langbroek, an archaeologist, astronomer and satellite tracker, of a portion of Geostationary Orbit. The photograph was taken using the SatTrackCam Leiden (Cospar 4353, formerly 4352), an amateur satellite tracking camera located in the Netherlands. The camera is used to make accurate positional measurements in

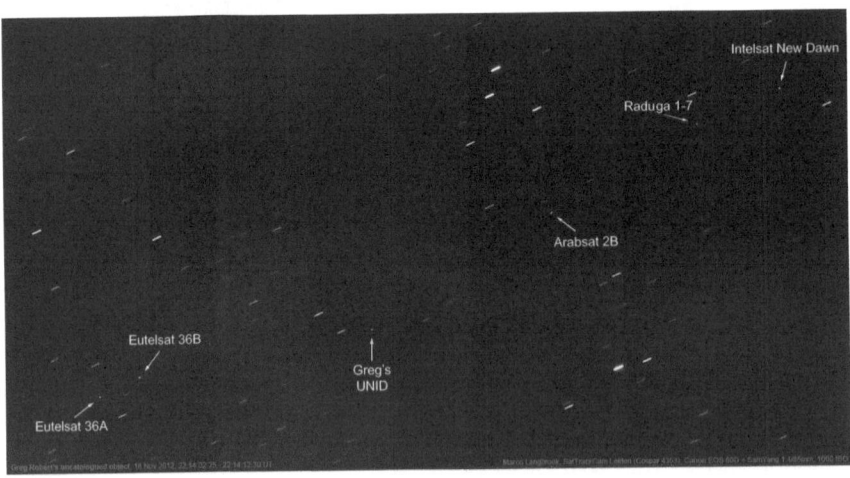

Fig. 3.3 Six geostationary objects were captured in this photograph and identified using published tracking data. Their location spans from 36° East (Eutelsat 36a) to 32.9° East (Intelsat New Dawn) (photo by and courtesy of Marco Langbroek)

order to determine satellite orbits. The image shows seven objects, distinguished from the background stars by being points rather than lines. There is no detail; the image shows only location in relation to the background stars and the orbital plane from the geocentric perspective.

Eutelsat 36A and 36B

In the lower left hand corner of the image are the co-located *Eutelsat 36A* and *Eutelsat 36B*, components of a telecommunications constellation owned by French-based satellite company Eutelsat SA. The company was established in 1977, with its first satellite launch in 1983. The dramatic events of 1989 created an impetus to expand:

> Geopolitical changes after the fall of the Berlin Wall transformed Eutelsat's original scope to serve customers in Western Europe. The organization's membership progressively expanded from the early 1990s to include all Eastern European countries (Eutelsat n.d. a).

Effectively, the expansion of the satellite fleet from four (1977–1989) to 11 in the following decade represents the expansion of capitalism into previously closed markets. Eutelsat currently has 34 satellites in GEO, supplying television, radio, and broadband internet. From 2012, the previous diversity of satellite names were rationalised by each being given a Eutelsat designation.

Eutelsat 36A is aimed at television broadcasting to Eastern Europe, Russia and the Commonwealth of Independent States (CIS), and sub-Saharan Africa, including Nigeria and neighboring countries (Eutelsat n.d. a), utilizing the Ku-band. The satellite was constructed by Alcatel Space Industries (now incorporated into Thales Alenia Space) and launched in 2000 on a Lockheed Martin Atlas IIIA rocket from Cape Canaveral. *Eutelsat 36B* provides television broadcasting, digital video and telecommunications in the Ku-band for Europe, Russia, Central Asia, and Africa. The satellite was constructed by Thales Alenia Space and launched on a Russian Proton Breeze M from Baikonur in Kazakhstan in 2009 (Eutelsat n.d. b).

ArabSat 2B

ArabSat is an Inter-Governmental Organization founded by the Arab League in 1976. Of the 21 member states across Africa and Middle East, Saudi Arabia is the major investor. The mission of ArabSat is to "Connect Arab societies and the world by providing reliable telecommunications services in harmony with Arab values and culture" (ArabSat 2013). ArabSat currently operates five satellites, at three orbital positions, including *ArabSat 2B*, (operating in the Ku-band and C-band), providing broadcast services, telecommunications and internet. The satellite was manufactured by Aérospatiale, a French-based manufacturing firm, and launched on an Arianespace Ariane 4 from Kourou in French Guiana in 1996 (Krebs 1996–2013a). Aérospatiale's satellite division merged with Alcatel to become Alcatel Space in 1999; the company was then taken over by Thales Alenia Space.

ArabSat 2B played an important role in the Arab Spring, a wave of anti-government protests and political demonstrations that overtook the Arab world starting in 2010. Despite the highly publicized role of social media, satellite broadcasting was in fact more critical in providing accessible coverage of the events and shaping opinion (Amin 2013). Satellite news channels such as Al Jazeera, Al Arabiya, the BBC's Arabic news, and France 24 were active in "giving a voice to the voiceless, covering opposition groups, exposing corruption, reporting demonstrations, and discussing issues of freedom, democracy, and social justice in the Arab states" (Amin 2013).

Raduga 1-7

The Raduga family, the USSR's first GEO satellites launched from 1974, was developed by NPO PM (later ISS Reshetnev) in Zheleznogorsk and constructed by PO Polyot in Omsk (Zak 2013). They were dual use, i.e., used for both civilian and military telecommunications, including civilian television and telephone.

The Raduga 1 series, however, launched from 1989, seems to have had only military applications. *Raduga 1-7* was launched in 2004 on a Proton K rocket from Baikonur. The spacecraft is registered with the United Nations as *Globus 1*. In February 2010, *Raduga-1-7* started to drift. Station-keeping maneuvers were performed, but in February 2011, the satellite appeared to have failed and was drifting westwards.

In early March 2011, a South Korean and two Japanese satellites were forced to undertake evasive maneuvers to avoid possible collisions (Zak 2013; see also Lee et al. 2012). Because of its military nature, there is limited information about this satellite.

Intelsat New Dawn

Intelsat SA has a venerable history. It was initially established as an intergovernmental consortium in 1964 in order to build the first commercial global satellite communications system. A privatized commercial service provider since 2001, it operates over 50 telecommunications satellites (Intelsat 2014). The *Intelsat 28 (New Dawn)* satellite was constructed by Orbital Sciences Corporation, one of ten GEOStar satellites commissioned by Intelsat, and launched on an Ariane 5 in 2011 from Kourou, French Guiana. It is the first African private sector communications satellite, a joint venture between a South African investor group Converge Partners and Intelsat SA to deliver internet and television to Africa (Krebs 1996–2013b; Orbital Sciences Corporation 2013).

The satellite was to have broadcast on C-Band and Ku-Band. When it reached GEO the C-band reflector antenna's deployment mechanism got caught in the insulation and failed to deploy. The Ku-band antenna was also caught in the billows, but was freed. Efforts to shake the C-band antenna loose were unsuccessful and cost a year's worth of fuel (Krebs 1996–2013b). It remains in GEO, with half its functionality and a reduced mission life.

Unknown 20121117

Langbroek's image shows a final object with an intriguing designation: Greg Robert's UNID, or the uncatalogued object known as *Unknown 20121117*. This object had first been observed by Greg Roberts in South Africa on November 17, and subsequently by Langbroek on November 18, during which time it appeared to have drifted west. There are more than 400 satellites drifting through GEO, and many more uncatalogued objects than there are catalogued (Jehn et al. 2005). Langbroek (2012) implies that it may be a military satellite.

Another possible source of uncatalogued objects is thermal blankets or multilayer insulation: these are known to have torn off LEO satellites, and Jehn et al. (2005: 1326) suggest that that they could account for objects visible from Earth in GEO, presumably because their cross-sectional areas are comparatively large. Of our small sample, this object is the only one which may fall into the standard definition of 'junk'.

From the Visible to the Intangible

Within this snapshot, numerous threads can be discerned. The location of the satellites is a function of their intended markets, and there is some degree of overlap, with, for example, *Eutelsat* and *Raduga* both covering Russia, while various parts of Africa are covered by *ArabSat 2B*, *Intelsat New Dawn*, and *Eutelsat 36A* and *36B*. This is a predominantly northern hemisphere reflection. The terms of the Outer Space Treaty of 1967 prevent national sovereignty being extended to space, and orbiting spacecraft are the ultimate disdainers of national boundaries. The impacts of this can be seen in the way satellite TV has united diaspora populations. For example, the broadcast of Arabic language television channels through *ArabSat 2A*, *ArabSat 2B,* and *Eutelsat* satellites has led directly to an expansion of Arab languages (Laroussi 2003: 251). More than that, Miladi (2006) argues that pan-Arab satellite networks have led to the emergence of a transnational public sphere. In the case of the collapse of communism in Eastern Europe, Eutelsat satellites were active participants a recolonization of capitalism across the former

'Iron Curtain'. The nature of orbital space and the overflight of the Earth under-
mine the coherence of nation-states back on Earth (Gorman 2009d).

The manufacture and launch of these satellites also runs counter to the strong
nationalist foundations of space industry arising from the Cold War. We see manu-
facturers emerging and coalescing (the incorporation of Alcatel and Aérospatiale
into Thales Alenia), and launches across three continents and former ideological
barriers. The consortia financing these satellites and the end users are multi- and
transnational, but manufacturing and launch are still provided by a limited number
of players.

This is also a very modern landscape, spanning 1996 (*ArabSat 2B*) to 2011
(*Intelsat New Dawn*). It represents a mature telecommunications industry rather
than the more tentative origins represented by *Syncom 3*, launched in 1963 (the
first GEO satellite) and *Early Bird*, launched by Intelsat in 1965. Presumably
such early satellites now form part of the drifting population outside the frame
of images like this because station-keeping is no longer performed to keep them
on-orbit. Also missing is the 'dark matter' of orbit, the fragments and objects too
small to be detected. Our perception of this part of space is very much a function
of the observation technology.

Eutelsat 36A and *36B* are part of a constellation of satellites, coordinated to
provide specific areas of coverage. This is a 'horizontal' spatial relationship. The
numbers assigned to all these satellites, however, indicate that that they are part
of a 'vertical' chronological relationship to the earlier incarnations, which did not
necessarily have the same form or function. A large number of these will be intact
among the catalogued population, even if no longer operational.

Using a photograph of this nature is useful mostly as a heuristic device to highlight certain objects and relationships. We can see the links between what appears
first to be a disparate and random grouping, and surmise deeper connections to
objects out of the shot. There is still an element missing, however. This visual
snapshot of the human hardware in geostationary orbit shows us only the material
component of a constant electromagnetic exchange with Earth.

Electromagnetic Spectrum: Beyond the Visual

The electromagnetic spectrum consists of all wavelengths of electromagnetic
radiation, including radio waves, microwaves, infrared, visible light, ultraviolet, gamma rays, and X-rays, with wavelengths from the photonic scale to metres
long (RF). The visual is only one way to understand the orbital landscape, one
that we naturally privilege by virtue of our confinement to a human sensorium.
Like orbits, however, the human use of spectrum is not evenly distributed, being
concentrated in certain high-density regions, particularly the microwaves used
for telecommunications. The spectrum is divided into 'bands', the commercial
and scientific use of which is regulated by the International Telecommunications
Union. Spectrum is considered to be a limited resource: if too many sources are

transmitting in a certain wavelength, the waves will interfere with each other and the encoded information is difficult to recover. In this case the band is said to be 'full' or 'congested'.

Within the 'natural' space environment, the Sun emits radiation across the entire spectrum, with predominance in the ultraviolet, visible and infrared regions, and planets like Jupiter are sources of radio waves. By contrast, terrestrial telecommunications are focused on the microwave frequencies. Most of the satellites discussed above utilise the Ku-band, from 12–18 GHz (the 'u' refers to the fact that it occurs *under* the K-band).

In the early days of satellite telecommunications, the C-band was used in combination with large antennas to provide communications across wide areas (NewSat nd). The growth of the industry saw the C-band increasingly taken up and under pressure from terrestrial encroachment. In the early 1990s, the satellite industry moved into the Ku-band, which utilised smaller antennas at both in space and on the ground (NewSat nd). The choice of these frequencies relates to the low level of signal loss through the atmosphere and the ability of antennas at the receiving end to 'gain' and amplify the weak signal, which has travelled 36,000 km from GEO.

The hardware exists only to receive and transmit signals, which are invisible and inaudible to us until reconfigured by signal capturing, processing and converting technology. Spectrum is both a driver of satellite telecommunications technology and an invisible 'soup' in which the spacecraft swim. This makes it very different from sensory landscapes of human interaction, composed of visible light wavelengths, sound, and the molecular interactions of smell and touch. It is truly non-human and robotic; our interaction with it can only be mediated by antennas and signal processors.

Of course, only satellites with enough power can transmit; the spectrum landscape is the land of the living. Satellite death occurs when the batteries fail and the transponders end transmission, which may take place after terrestrial antennas have ceased to listen (when a tree falls in the forest ...). The electromagnetic footprint of orbital objects is crystallized in the form of their antennas, and in data storage within the spacecraft and back on Earth.

Defunct satellites could be compared to cranial endocasts in the study of language origins: the internal structure can be discerned, but the transmissions are as ephemeral as speech.

Emergent Properties of the Orbital Robot Assemblage

In the preceding sections, I have mapped out features of GEO based on a visual slice framed by Langbroek's photograph and the electromagnetic environment, to make some sense of this landscape. Now, with reference to Deleuze and Guattaristyle assemblages, I propose to leave the recent past behind and venture into the future. The emergent properties of our robot avatars are worth considering as a way of distinguishing what is relevant to research or heritage in the present.

Currently, we can consider satellites to be robots with a high 'neglect tolerance' (Goodrich and Schultz 2007: 239). It is a rare satellite indeed that attracts direct human intervention on-orbit—the Hubble Space Telescope in LEO (serviced five times) is perhaps the only one. But higher degrees of autonomy and decision-making are being planned in the form of self-organizing swarms of satellites (e.g., Bonnet and Tessier 2007; Izzo and Petazzi 2007). The swarm is a classic assemblage, creating a coherent pattern through the accumulation of individual movements.

At the same time as this proposed evolution of orbital objects, entropy may be on the verge of taking over the system. The so-called Kessler Syndrome is an emergence that could occur in the near future. In popular (and occasionally scientific) contexts, the Kessler Syndrome is the worst case scenario for space junk: a cascade of random collisions that create so much debris the Earth is enveloped and cut off from space. When Fengyun 1C was destroyed, many feared it would be the 'tipping point' into the cascade. However, this conception is not strictly accurate.

The Kessler Syndrome derives from an early paper by Kessler and Cour-Palais (1978), in which they argued that a situation could arise in which "the debris flux will increase exponentially with time, even though a zero net input may be maintained" (Kessler et al. 2010: 47). The idea of such a cascade derives from planetary formation, where collisions cause the lowering of inclination and the eventual formation of a ring or belt around a celestial body. However, LEO debris collisions are highly unlikely to behave like other planetary rings, as the smaller particles will be removed by atmospheric drag before that point. While Kessler et al. (2010) argue that the orbital debris situation is indeed critical, they stop short of endorsing the irreversible negative feedback version of the Syndrome.

Nonetheless, a Kessler-type situation takes this assemblage to a new level of increasing complexity, and evokes a concept more at home in the realm of the Search for Extraterrestrial Intelligence (SETI). In 1960, physicist Freeman Dyson discussed dismantling Jupiter to create a shell around the sun for the purposes of harvesting solar energy (Dyson 1960). A modest sphere might circle the sun; on a more substantial scale, the sphere might enclose both the sun and the planet, or even a whole solar system. The Dyson sphere would maximize energy harvesting to support a 'civilization' with far greater requirements than our own. Although frequently imagined as a rigid, solid sphere, in fact Dyson postulated a swarm of objects designed to collect the light from the central star (Carrigan 2008: 19).

The identification of such megascale structures outside our solar system is part of SETI research programmes (e.g. Jugaku and Nishima 2000). We are far from this stage in our use of space, but it's hard not to see parallels in the shell our own planet is acquiring; and although satellites use their solar arrays solely for their own power, their capacity to harvest solar energy could be turned to other uses. The earliest version of such a structure might recycle material already in orbit, to minimize the great expense of launching raw materials. Later, asteroids and moons might be sourced for materials. In this trajectory, the next step is the computing requirements of a culture able to harness the totality of a sun's energy output. "Information processing superobjects" or Jupiter brains (Sandberg 1999) were the precursors to a much more ambitious feat of the imagination.

The Matrioshka Brain (MB) was conceived by Bradbury (1997–2000). The MB is a computer, constructed from the matter of an entire solar system, and operating from the nanoscale to the megascale. The 'thought' processes take place at scales beyond human engagement, at the nano- and solar system scale, and Bradbury considers that human interaction with an MB would be pointless. They would most probably be constructed in concentric shells around a star, like a set of nested Dyson spheres or swarms. The star provides the energy source and the waste heat radiated by the innermost shell is taken up by the next shell, and so on. The key components of the swarm are designed for power collection (e.g. mirrors or solar cells at their most simple), heat disposal, radiation protection, computing, and data storage.

Together, this structure makes up a 'computronium' element, and each level of shell has the computronium appropriate for its operating temperature, which decreases with distance from the sun. Although Bradbury does not say this, he has effectively described contemporary satellites, which must have all these components in order to function. As an assemblage, whole satellites in Earth orbit could be argued to constitute proto-computronium.

Conclusions

The material, mathematical and computational requirements of these megascale structures have been investigated (Bradbury 1997–2000; Sandberg 1999), but their existence is, of course, highly speculative. In drawing a thread that connects self-organizing satellite swarms, debris clouds, and Matrioshka brains, my intention is to recognize the transformative potential of orbital debris and satellites, as our robot avatars in space.

Archaeological insights into Earth orbit entail both looking back and looking forward. An archaeologist in 100 years may study the artifacts that are now on the cusp of the future and perceive connections that we cannot even dream about. But the 'foreseeable' future (Crowther 1994) is closer than we realize and it may be time to redefine space 'junk'. The 96 % of catalogued non-functioning orbital objects, from whole spacecraft to fragmentation debris, is a resource for future space industry.

An aggregate approach to orbital artifacts, which treats spatial associations as insignificant and focuses on individual spacecraft, is not sufficient to capture the dazzling complexity of the orbital assemblage. Landscape and assemblage offer avenues to focus on relationships between objects, and between objects and their environment. Thus, the proximity of the satellites to each other is related to the terrestrial footprint they service. These positions are maintained by station-keeping, which prevents the satellites from colliding (as nearly happened when *Raduga 1-7* drifted) and keeps them within reach of domestic antennas.

Their relationship within this part of GEO reflects changing terrestrial politics in the last three decades: the fall of European communism, the Arab Spring, the growth of a pan-Arab culture, and not least, the interweaving of military and security interests among the television broadcasts watched in living rooms. Cultural and geographical aspects of GEO have been explored by Collis (2009) and Parks (2005); the archaeological challenge is to make meaningful inferences about the assemblage of orbital space which takes into account the physicality of spacecraft, debris, and the electromagnetic environment.

What are sketched here are the rudiments of a larger project that should draw on the catalogue, and tools such as MASTER and STK. Visualizations of orbit are currently used as predictive tools in Space Situational Awareness, but this data can also be used to investigate historical and technological relationships. For example, aspects of the design of GEO satellites like *Eutelsat 36A, 36B* and *Intelsat New Dawn,* such as the classic solar array configuration, could be represented across space and time. The stylistic and technological variation between, say, scientific satellites manufactured within a national research program and those made in the transnational context of modern telecommunications, is the kind of question that has not yet been explored.

While it is important to draw space further into the sphere of human interaction, there is no avoiding the tendency of the cultural landscape to privilege the human, as we are the stakeholders interested in its heritage value. Posthuman perspectives (e.g. Plumwood 1996) shift the balance. Calling a satellite a robot rather than simply a machine emphasizes its autonomy, despite its lack of sentience. Satellites are more part of our world than most people are conscious of, but their adaptation to the space environment and physical distance from us creates an isolation and silence. We control them and yet they are beyond our control, as the orbital debris problem illustrates. We can only hear their voices and speak to them by translating their wavelengths into the range of human perception.

One day, they will be creating their own artifacts and trace fossils independently of human desires (Spennemann 2007). Here, in the early decades of this trajectory, an archaeological perspective connects the physicality of space junk to the space it occupies and to the unseen world of the full spectrum. Our robot avatars have already taken on a life of their own.

Acknowledgments I'd like to particularly thank Marco Langbroek for his awesome photography skills and permission to use the photograph. Much gratitude to Charles Stross, who inspired me to start looking at Matrioshka brains, and provided valuable research leads. I would also like to thank Twitter colleagues @spacearcheology and @JohnJRoby who assisted in tracing obscure references and terms, and @SarahMay_1 and @LornaRichardson for transcontinental writing company in #madwriting.

References

Amin, H. (2013). Role and impact of satellite broadcasting during the Arab Spring. Paper Presented at the Annual Meeting of the BEA, Las Vegas Hilton, Las Vegas, NV. Accessed December 13, 2013. http://citation.allacademic.com/meta/p545299_index.html

ArabSat. (2013). *Mission and vision.* Accessed August 15, 2013 http://www.arabsat.com/pages/AboutUs.aspx

Australia ICOMOS. (2013). *The Burra Charter. The Australia ICOMOS Charter for places of cultural significance.* Available at http://australia.icomos.org/publications/charters/

Belk, C. A., Robinson, J. H., Alexander, M. B., Cooke, W. J., & Pavelitz, S. D. (1997). *Meteoroids and orbital debris. Effects on spacecraft.* NASA Reference Publication 1408.

Bonnet, G., & Tessier, C. (2007). Collaboration among a satellite swarm. In *Proceedings of the 6th International Joint Conference on Autonomous Agents and Multiagent Systems.* Article No 54. New York: ACM.

Bradbury, R. (1952). A sound of thunder (pp. 20–21). Collier's, 28 June.

Bradbury, R. (1997–2000). *Matrioshka brains*, MS.

Capelotti, P. J. (2010). *The human archaeology of space. Lunar, planetary and interstellar relics of exploration.* Jefferson: McFarland and Company.

Carrigan Jr., R. A. (2008). *IRAS-based whole-sky upper limit on Dyson spheres.* Fermilab-pub008-352-AD.

Collis, C. (2009). The geostationary orbit: A critical legal geography of space's most valuable real estate. In D. Bell & M. Parker (Eds.), *Space travel and culture: From Apollo to space tourism* (pp. 47–65). Malden: Wiley-Blackwell/The Sociological Review.

Crowther, R. (1994). The trackable debris population in low Earth orbit. *Journal of the British Interplanetary Society, 47*(4), 28–133.

De Landa, M. (2006). *A new philosophy of society: Assemblage theory and social complexity.* London: Continuum.

De Landa, M. (2002). *Intensive science and virtual philosophy.* New York: Continuum.

Deleuze, G., & Guattari, F. (1987). *A thousand plateaus: Capitalism and schizophrenia.* Translated by Brian Massumi. Minneapolis: University of Minnesota Press.

Dyson, F. (1960). Search for artificial stellar sources of infrared radiation. *Science, 131*(3414), 1667–1668.

Eutelsat. n.d. (a). *Expanding reach and building a dynamic business.* Accessed November 18, 2013. http://www.eutelsat.com/en/group/our-history/1990-2001.html

Eutelsat. n.d. (b). *The Fleet 36° Eutelsat 36a.* Accessed November 18, 2013. http://www.eutelsat.com/en/satellites/the-fleet/EUTELSAT-36A.html

European Space Agency. (2013). *Space debris.* Accessed November 12, 2013. http://www.esa.int/Our_Activities/Operations/Space_Debris/About_space_debris

Flannery, Kent. (1968). Archaeological systems theory and early Mesoamerica. In B. J. Meggers (Ed.), *Anthropological archaeology in the Americas* (pp. 67–87). Washington: Anthropological Society of Washington.

Goodrich, M. A., & Schultz, A. (2007). Human–robot interaction: A survey. *Foundations and Trends in Human–Computer Interaction, 1*, 203–275.

Gorman, A. C. (2009a). The gravity of archaeology. *Archaeologies: The Journal of the World Archaeological Congress, 5*(2), 344–359.

Gorman, A. C. (2009b). The cultural landscape of space. In A. Darrin & B. L. O'Leary (Eds.), *The handbook of space engineering, archaeology and heritage* (pp. 331–342). Boca Raton: CRC Taylor and Francis Press.

Gorman, A. C. (2009c). The archaeology of space exploration. In D. Bell & M. Parker (Eds.), *Space travel and culture: From Apollo to space tourism* (pp. 129–142). Malden: Wiley-Blackwell/The Sociological Review.

Gorman, A. C. (2009d). Beyond the space race: The significance of space sites in a new global context. In A. Piccini & C. Holtorf (Eds.), *Contemporary archaeologies: Excavating now* (pp. 161–180). Bern: Peter Lang.

Gorman, A. C. (2005a). The cultural landscape of interplanetary space. *Journal of Social Archaeology, 5*(1), 85–107.

Gorman, A. C. (2005b). The archaeology of orbital space (pp. 338–357). In *Australian Space Science Conference 2005.* Melbourne: RMIT University.

Gorman, A. C., & O'Leary, B. L. (2007). An ideological vacuum: The cold war in space. In J. Schofield & W. Cocroft (Eds.), *A fearsome heritage: Diverse legacies of the cold war* (pp. 73–92). Walnut Creek: Left Coast Press.

Green, C., & Lomask, M. (1970). *Vanguard: A history*. Washington, DC: Scientific and Technical Information Division, National Aeronautics and Space Administration.

Hyde, J. L. (2000a). As-flown shuttle orbiter meteoroid/orbital debris assessment, Phase I, JSC-28768.

Hyde, J. L. (2000b). As-flown shuttle orbiter meteoroid/orbital debris assessment, Phase II, JSC-29070.

Intelsat. (2014). *Company facts*. Accessed January 3, 2014. http://www.intelsat.com/about-us/company-facts/

Izzo, D., & Pettazzi, L. (2007). Autonomous and distributed motion planning for satellite swarm. *Journal of Guidance, Control and Dynamics, 30*(2), 449–459.

Jehn, R., Agapov, V., & Hernández, C. (2005). The situation in the Geostationary ring. *Advances in Space Research, 35*, 1318–1327.

Johnston, E. (2013). *List of satellites in geostationary orbit*. Accessed January 8, 2014. http://www.satsig.net/sslist.htm

Jugaku, J., & Nishima, S. (2000). A search for Dyson spheres around late-type stars in the solar neighbour III. In G. Lemarchand, & K. Meech (Eds.) *Bioastronomy 99: A new era in the search for life* (pp. 581–584). ASP Conference Series, vol. 213.

Kessler, D. J., Johnson, N. L., Liou, J.-C., & Matney, M. (2010). The Kessler Syndrome: Implications to future space operations. *Advances in the Astronautical Sciences, 137*, 47–61.

Kessler, D. J., & Cour-Palais, B. G. (1978). Collision frequency of artificial satellites: The creation of a debris belt. *Journal of Geophysical Research, 83*(A6), 2637–2646.

Krebs, G. (1996–2013a). *Arabsat 2A, 2B*. Gunter's space pages. Accessed November 18, 2013. http://space.skyrocket.de/doc_sdat/arabsat-2a.htm

Krebs, G. (1996–2013b). *New Dawn → Intelsat 28*. Accessed November 18, 2013. http://space.skyrocket.de/doc_sdat/new-dawn.htm

Laroussi, F. (2003). Arabic and the new technologies. In J. Maurais & M. A. Morris (Eds.), *Languages in a globalising world* (pp. 250–259). Cambridge: Cambridge University Press.

Langbroek, M. (2012). *PAN, other Geostationary satellites, and another UNID* (*this time Greg's*). Accessed August 21, 2013. http://sattrackcam.blogspot.com.au/

Lee, B.-S., Yoola, H., Kim, H.-Y., & Kim, B.-Y. (2012). *GEO satellite collision avoidance maneuver strategy against inclined GSO satellite*. Paper presented at SpaceOps12, Stockholm.

Levine, A. S. (Ed.) (1991). *LDEF, 69 months in space: First post-retrieval symposium*. Washington, D.C.: National Aeronautics and Space Administration, Office of Management, Scientific and Technical Information Program.

Liou, J.-C., & Anz-Meador, P. (2010). An analysis of recent major breakups in the low Earth orbit region. Paper number IAC-10,A6,2,13,x6484. Accessed January 10, 2014. http://www.iafastro.net/iac/archive/browse/IAC-10/A6/2/6484/

McCray, P. W. (2008). *Keep watching the skies!: The story of Operation Moonwatch and the dawn of the space age*. Princeton: Princeton University Press.

Marcus, G. E., & Saka, E. (2006). Assemblage. *Theory, Culture and Society, 23*(2–3), 101–109.

Mehrholz, D., Leushacke, L., Flury, W., Jehn, R., Klinkrad, H., & Landgraf, M. (2002). Detecting, tracking and imaging space debris. *ESA Bulletin, 109*, 128–134.

Miladi, N. (2006). Satellite TV news and the Arab diaspora in Britain: Comparing Al-Jazeera, the BBC and CNN. *Journal of Ethnic and Migration Studies, 32*(6), 947–960.

NASA Orbital Debris Program Office. (2012). *Orbital debris frequently asked questions*. Accessed August 1, 2013. http://orbitaldebris.jsc.nasa.gov/faqs.html#3

NASA Marshall Space Flight Center. (2008). *Advanced space transportation program: Paving the highway to space*. Accessed August 18, 2013. http://www.nasa.gov/centers/marshall/news/background/facts/astp.html

NewSat. nd. *Ku-band capacity*. Accessed January 3, 2014. http://www.newsat.com/Satellites/ku band.html

Orbital Sciences Corporation. (2013). *Intelsat New Dawn fact sheet*. Accessed January 10, 2014 www.orbital.com/newsinfo/publications/NewDawn_Fact.pdf

Parks, L. (2005). *Cultures in orbit: Satellites and the televisual*. Durham: Duke University Press.

Pardini, C., & Anselmo, L. (2009). Assessment of the consequences of the Fengyun-1C breakup in Low Earth Orbit. *Advances in Space Research, 44*, 545–557.

Pardini, C., & Anselmo, L. (2011). Fengyun 1C diagram physical properties and long-term evolution of the debris clouds produced by two catastrophic collisions in Earth orbit. *Advances in Space Research, 48*(3), 557–569.

Phillips, J. (2006). Agencement/assemblage. *Theory, Culture and Society, 23*(2–3), 108–109.

Plumwood, V. (1996). Androcentrism and anthrocentrism: Parallels and politics. *Ethics and the Environment, 1996*, 119–152.

Prigogine, I., & Stengers, I. (1984). *Order out of chaos. Man's new dialogue with nature*. Toronto: Bantam Books.

Salmon, M. (1978). What can systems theory do for archaeology? *American Antiquity, 43*(2), 174–183.

Sandberg, A. (1999). The physics of information processing superobjects: Daily life among the Jupiter brains. *Journal of Evolution and Technology, 5*(1), 1–34.

Schiffer, M. B. (2013). *The archaeology of science. Studying the creation of useful knowledge*. Heidelberg: Springer.

Spennemann, D. H. R. (2007). Of great apes and robots: Considering the future(s) of cultural heritage. *Futures, 39*, 861–877.

Venn, C. (2006). A note on assemblage. *Theory, Culture and Society, 23*(203), 107–108.

Zak, A. (2013). *Raduga 1*. Accessed August 18, 2013. http://www.russianspaceweb.com/raduga 1.html

Chapter 4
Mobile Artifacts in the Solar System and Beyond

P.J. Capelotti

Abstract This chapter looks at the discard behavior by humans and the ability of space technology to have a life history as artifacts become archaeological objects and enter the archaeological record. Once a spacecraft, for example, no longer responds to signals from Earth and ceases to be used for which it was designed, it becomes a discarded, and hence, archaeological object. The author explores both classical and innovative methods and provides historic case studies of spacecraft to illustrate how archaeologists can study this class of artifacts.

Introduction

On 13 September 2013, the National Aeronautics and Space Administration announced that *Voyager 1* had entered a region of space characterized by cold, dense plasma, an area qualitatively different from the solar plasma the spacecraft had sailed through since its launch in 1977 (NASA 2013). The changes in the contextual surround of the artifact had first been detected 13 months earlier, and the new data confirmed what had been expected since that time: *Voyager 1* had crossed the heliopause, the theoretical boundary where the solar wind from the Sun is deflected by the denser medium of interstellar space, and had become the first artifact made by humans to leave the Solar System.

Designed to explore the outer planets of the Solar System, *Voyager 1* and its twin, *Voyager 2*, then continued their respective flights and are now the longest, continuously operated spacecraft in the human exploration of space. While the news that humans had designed and constructed an artifact capable of leaving the Solar System while still in systemic context cheered scientists who sought

P.J. Capelotti (✉)
Division of Social Sciences, Abington College, Penn State University, Abington, PA, USA
e mail: pjc12@psu.edu

© Springer International Publishing Switzerland 2015
B.L. O'Leary and P.J. Capelotti (eds.), *Archaeology and Heritage of the Human Movement into Space*, Space and Society, DOI 10.1007/978-3-319-07866-3_4

Fig. 4.1 This artist's concept shows the general locations of NASA's two Voyager spacecraft. Voyager 1 (*top*) has sailed beyond our solar bubble into interstellar space, the space between stars. Its environment still feels the solar influence. Voyager 2 (*bottom*) is still exploring the outer layer of the solar bubble (*source* NASA/JPL—Caltech)

to gather data on the nature of interstellar space, it also begged the question of how long the remarkable artifact would continue to broadcast its reports from the newly-entered territory. *Voyager 1* communicates daily with its originating planet through signals emitted at about 23 W, "the power of a refrigerator light bulb" (NASA 2013). Yet even this minute output is expected to cease in the year 2020, when *Voyager 1* will transition from its current systemic context to its permanent status as an inert artifact in a new, archaeological, context (Fig. 4.1).

Aerospace and Industrial Archaeologies

Pedagogically, the study of spacecraft and the launch facilities required to operate them as archaeological objects involves the application of well-established forms. The overall approach owes much to the field of industrial archaeology (Raistrick 1972), and seldom has an archaeological perspective on the remains of the space age been better described than by Steinberg (2000: 104), who studied the abandoned launch complexes at Cape Canaveral, Florida, and found that at the "cape where it began, the past has already begun already. The early sites of American space travel present themselves as crumbling ensembles of concrete that are reverting back to nature…".

Fig. 4.2 The Apollo 12 lunar module touched down as planned within about 180 m of the Surveyor 3 probe. The probe had landed on the Moon on 20 April 1967 in the edge of a small crater and returned images of the lunar surface. On the second moonwalk EVA on 20 November 1969 astronauts Conrad and Bean visited Surveyor 3, inspected the craft, and removed some parts for return to Earth. Here Conrad is shown examining the camera on Surveyor 3. The lunar module can be seen in the background in this northwest looking photo taken by Alan Bean. The Surveyor camera was one of the parts returned and is on exhibit at the National Air and Space Museum in Washington, DC (*source* NSSDC, Apollo 12, AS12-H-48-7133)

> The foundations and concrete structures are gradually covered by sand. Grass is growing on the underground control rooms. The spherical tanks for the liquid rocket fuel loom like mythological reliquaries into the world of the lagoons.

While it seems rather extraordinary to consider the possibilities of archaeological field research on artifacts of the Space Age discarded on lunar or planetary bodies other than Earth, we can take note of the fact that it has already taken place. In fact, NASA has been conducting formational archaeology research on lunar sites at least since 1969 (Nickle 1971: 2683–2697; NASA 1972). On 20 November 1969, astronaut Charles Conrad, Jr., recovered pieces (the television camera, remote sampling arm, and pieces of tubing) from the unmanned *Surveyor III* probe that had soft-landed in the Moon's Ocean of Storms on 19 April 1967 (Fig. 4.2). These artifacts were brought back to Earth so that the Johnson Space Center in Houston, Texas, and the Hughes Air and Space Corporation in El Segundo, California, could analyze the natural transformational processes operating on aerospace artifacts left on the Moon.

In the particular cases of mobile remnants of the human exploration of the Solar System and beyond, these artifacts possess the significant difference that such artifacts—barring inflight collision or atmospheric disintegration—will remain indefinitely in motion and, especially when compared to the ruins of Cape Canaveral, in a state of relatively perfect preservation. For example, once it becomes "strictly archaeological" (see below) after 2020, *Voyager 1* could survive long after the extinction of intelligent life itself.

There are contradictions for the archaeologist to consider here, but theoretically they are hardly insurmountable. It is without doubt a bit odd for an archaeologist to consider objects that are still in motion as part of the material culture database, since the image most have of archaeological research centers upon the careful excavation of artifacts or fossils fixed within the soil or rock of the Earth, some for millions of years. New categories of archaeological method and theory seem to be called for if we are to consider the possibilities of fieldwork on the now dead—or soon-to-be dead—spacecraft launched into space, like *Voyager 1*, that might themselves afloat amid interstellar space for millions of years.

The Theory of Aerospace Archaeology

Theoretically, we choose to employ, in Dymond's (1974: 7) words, "the widest-possible definition of archaeology [with] the use of a great range of documents." In this manner, the field of aerospace archaeology traces its intellectual underpinnings to the work of pioneer archaeologist O.G.S. Crawford and anthropologists Richard A. Gould and Ben R. Finney, the latter of whom wrote extensively on the notion of humans as a species that "evolved as an exploratory, migratory animal" (Finney 1992: 105).

In the 1980s, in a tangential relation to his work on the technology of Polynesian seafaring and its bearing on human migration and exploration, Finney began to study an "expansionary phase that was not only of potentially greater import than any terrestrial example, but one that could be studied directly without having to reconstruct and test the vehicles involved or interpret ambiguous texts" (Finney 1992: 2). Finney was referring to human expansion into space, and his thesis of a human species as an essentially exploratory one comprises the cornerstone of archaeological inquiries into the human and robotic exploration of space.

While O.G.S. Crawford "referred to obsolete aircraft as strictly archaeological" (Dymond 1974), and James Deetz referred to "interplanetary space vehicles" as a complex example of material culture [in Deetz's words: "that sector of our physical environment that we modify through culturally determined behavior" (Deetz 1977: 24)], the first suggestion that aircraft wrecks might yield important anthropological data was made as early as 1983 by Richard A. Gould. Gould suggested that debates originating in the historical record could be evaluated through "the explanatory potential of archaeology… [where] differing historical interpretations can be regarded as a source of alternate hypotheses, with archaeological evidence being used to test each alternative" (Gould 1983: 117–118).

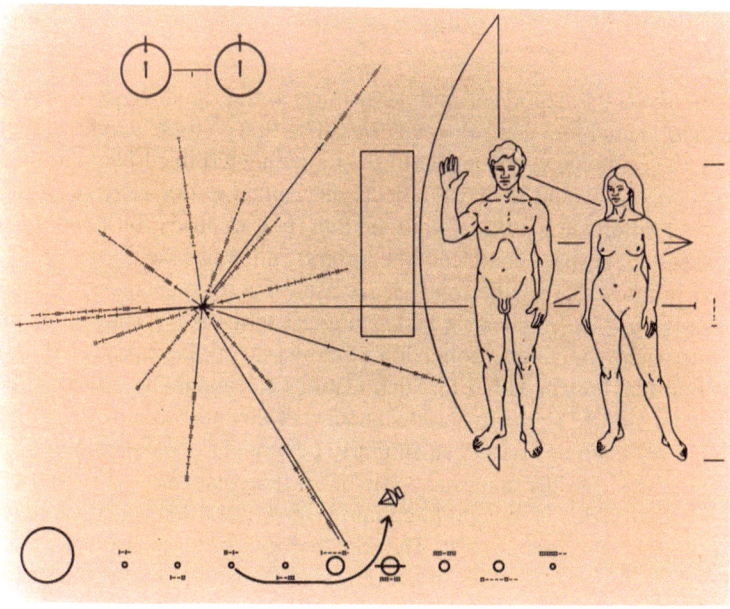

Fig. 4.3 The Pioneer 10 plaque features a design engraved into a gold-anodized aluminum plate, 152 by 229 mm (6 by 9 inches), attached to the spacecraft's antenna support struts to help shield it from erosion by interstellar dust (*source* Fimmel and Swindell 1977: 182)

There is first the issue of whether or not a mobile artifact is in fact archaeo-logical. Schiffer's definition rests on whether or not an artifact is still being used for that which it was designed. Once a spacecraft no longer responds to signals from Earth, it ceases to be used for the original mission for which it was designed, and becomes instead a discarded, and hence, archaeological, object (for a com-plete discussion of these categories, see Schiffer 1987). This is the case with the *Pioneer 10* space probe, which ceased 'speaking' with Earth in 2003, and is now headed on a two million year journey toward the red star Aldebaran (although, significantly, it can be interpreted as having left its active, systemic context as a mobile collector of scientific data to a new systemic context as an emissary of Earth, since it carried on board the famous Pioneer plaque, meant as a message of greeting to other forms of intelligence in the universe: Fig. 4.3).

As we see in the example of *Pioneer 10*, mobile artifacts—as for more con-ventionally-considered artifacts on Earth—the strict categories of systemic and archaeological context for artifacts in space are not absolute. It is possible for an object to move in and out of context. By these lights, several distant-travelling spacecraft have not yet become strictly archaeological objects. These include the two Voyager probes, the Galileo probe to Jupiter and the New Horizons probe to Pluto. Each of these spacecraft is scheduled to break contact with Earth within the next decade, at which point, by definition, they will transition to archaeological objects. Until then, they remain a vital part of a living human cultural system.

Mobile Artifacts in the Solar System

Within the general category of mobile artifacts in the Solar System and, eventually, beyond, there are a few sub-categories. The first of these would be those artifacts that are on their way from Earth to some undetermined place in interstellar space; the second would be those objects that, either deliberately or as the result of a mission failure, now orbit around the Sun (heliocentric orbit); and, finally, the vast archaeological space 'midden' that currently encircles the Earth.

For our purposes, we will largely leave aside the third category, but here offer a few words on it. A 'midden' in archaeological terms is a feature of an archaeological site where one finds a collection of the waste products produced during the course of normal human daily life. Such features can and do accumulate for generations and be studied as the material signature of an entire culture. The midden of space junk that encircles Earth can similarly be thought of as a dump for domestic waste, but in this case the domestic waste of a whole planetary community.

The orbital midden is thought to comprise tens of millions of separate artifacts, none of them in systemic context. The vast majority of these artifacts are chips of paint from orbital satellites and spacecraft, slag from solid rocket motors, coolant from nuclear power plants and other such small debris. Some of this material will eventually fall from its Earth orbit and burn up in the atmosphere. But collisions between these small fragments create more small fragments, an increasing problem for attempts to track these objects so that they do no damage to new space missions.

The second category of mobile artifacts in the Solar System is those in heliocentric orbit, that is, orbit around the Sun. There are more than fifty such objects, including twenty-nine from launched from the United States, fifteen from the Soviet Union/Russian Federation, five from the European Space Agency, and five from Japan (for a complete list, see Capelotti 2010: 165–166). Only seven of these missions is still active and therefore in systemic context. The remainders are artifacts that were either lost through communications failures or technical problems, or those deliberately abandoned after their system use was finished—such as the Apollo 10 Lunar Module *Snoopy*—or those that missed their original targets such as the Moon or Venus or Mercury and were subsequently captured by the gravitational pull of the Sun.

The first category of mobile artifacts in the Solar System is that of those objects that were deliberately launched from Earth onto a journey of exploration of the Solar System that would continue with the spacecraft venturing towards interstellar space. These special artifacts of human intelligence hold a particular fascination for archaeologists, as they represent an attempt by *Homo sapiens* to fashion a tool that ultimately can cross the barrier that separates the Solar System from interstellar space on a hopeful mission to communicate with other forms of intelligent life that may or may not exist elsewhere in the galaxy. There are five such composite tools: *Pioneer 10*, *Pioneer 11*, *Voyager 1*, *Voyager 2*, and *New Horizons*. Each of these spacecraft initially had a specific scientific mission to carry out within the Solar System (*New Horizons* is still in its active mission phase; it will reach its target of the planet Pluto in the year 2015).

Once these primary missions were concluded, the spacecraft were then directed toward the boundaries of the Solar System with the expectation that they would eventually—as *Voyager 1* has now done— enter interstellar space and become the representatives of *Homo sapiens* to the rest of the galaxy.

Pioneer 10

The most famous of these representatives is perhaps *Pioneer 10* and its associated plaque. In a NASA history of the Pioneer expeditions, there is an epilog titled "Interstellar Cave Painting" (see Fimmel and Swindell 1977). It recounts how Eric Burgess of the *Christian Science Monitor* looked at the *Pioneer 10* spacecraft as it was being tested and conceptualized it as the first human object that might come into contact with other intelligent species. This observation led to a chain of events that culminated in the design and placement on the spacecraft of "a special message from mankind" (Fimmel and Swindell 1977: 183). Astronomers Carl Sagan and Frank Drake designed a plaque that would attempt in symbols to represent where the object had come from and the kinds of beings who had constructed it and sent it into space. While it is a brilliant work of human optimism, there are unfortunately no other representations of any other life on Earth, present or past, or any suggestion that the form of human on the plaque was the product of millions of years of evolution through natural selection and therefore constantly changing along with all other forms of life on the planet and, in fact, the planet itself.

As tool-using, bipedal organisms with stereoscopic vision, it often seems as if we have fashioned our space probes almost strictly for scientific purposes. When there is a thought given to cultural or archaeological implications of our exploration of space, as with the *Pioneer 10* plaque, the result can be seen as almost cursory—or illusory, depending on your interpretation of the *Pioneer 10* plaque. We are asking for an intelligence elsewhere in the Galaxy to possess the ability to detect our lifeless probe and the technical skill to intercept it as it moves at perhaps 10 km/s or faster, combined with the intellect to decipher and, potentially, to seek out the creators of such an unusual object. Yet the Pioneer plaque at least accomplishes the critical mission of letting others—as well as ourselves—know that *Homo sapiens* seek to learn from extraterrestrial civilizations and not just act of processors of positive data collected by our machines.

Assuming that the *Pioneer 10* probe and its unique representation of *Homo sapiens* are not destroyed by another species possessing some of our own aggressive tendencies, it should survive well beyond the age of humans on Earth. More than that, as the NASA history puts it, this modern cave painting "might survive not only all the caves of Earth, but also the Solar System itself. It is an interstellar stela that shows mankind possesses a spiritual insight beyond the material problems of the age of human emergence" (Fimmel and Swindell 1977: 185).

This brings us finally to the moment of transition between systemic and archaeological. Almost by definition, much of the history of aerospace exploration is a process of rolling discard and a decided lack of recycling. The Saturn V is the most obvious example of designed stages of discard, but virtually every aerospace artifact exhibits this model (the author once sat in a fascinating lecture by the late physicist Gerard K. O'Neill where he complained bitterly about the lost opportunities for constructing space colonies from all the materials NASA routinely discarded—rocket stages, fuel cells, even entire spacecraft).

As for the transition of *Pioneer 10* from its systemic to its current archaeological context, the details of this transition reveal the complex nature of archaeological formation processes (for a complete archive of this mission, see the NSSDC). The different components of the probe failed or were turned off at various times over more than 30 years. Some were turned off permanently and others were cycled on and off in accordance with a power sharing plan implemented in September 1989. The Asteroid/Meteoroid Detector failed in December 1973, followed by the Helium Vector Magnetometer (HVM) in November 1975 and the Infrared Radiometer in January 1974. The Meteoroid Detector was turned off in October 1980 due to inactive sensors at low temperatures. The spacecraft sun sensors became inoperative in May 1986, and the Imaging Photopolarimeter (IPP) instrument was used to obtain roll phase and spin period information until being turned off in October 1993 to conserve power. The Trapped Radiation Detector (TRD) and Plasma Analyzer (PA) were respectively turned off in November 1993 and September 1995 for the same reason. As of January 1996 the final power cycling plan included part-time operations of the Charged Particle Instrument (CPI), the Cosmic Ray Telescope (CRT), the Geiger Tube Telescope (GTT), and the Ultraviolet Photometer (UV). As of August 2000, only the GTT instrument was still returning data.

Humans ceased routine tracking of the probe on 31 March 1997 when they could not afford to pay for it. The last successful data acquisitions through NASA's Deep Space Network (DSN) occurred on 3 March 2002, the 30th anniversary of *Pioneer 10's* launch date, and on 27 April 2002. The spacecraft signal was last detected on 23 January 2003. No signal at all was detected during a final attempt to communicate with the probe on 6/7 February 2003, so Pioneer Project staff at NASA Ames concluded that the spacecraft power level had fallen below that needed to power the onboard transmitter, so no further attempts would be made. It can be assumed that, soon after that moment, the spacecraft entered its archaeological context, where it will permanently remain unless found and reused or recycled by another civilization.

Voyager 1

A similar and equally instructive history obtains for *Voyager 1*. Each Voyager spacecraft had mounted to one of the sides of the bus a twelve-inch gold-plated copper disk. Unlike the simple representation of the naked *Homo sapiens* on the Pioneer plaque, the Voyager disk had recorded on it a wide variety of sounds and images from Earth that were designed to portray the diversity of life and culture on the planet. Each disk is encased in a protective aluminum jacket along with a cartridge and a needle, which was the prevailing technology for playing recorded music at the time the Voyager spacecraft were designed and built. Instructions explaining from where the spacecraft originated and how to play the disk are engraved onto the jacket.

Electroplated onto a 2 cm area on the cover is also an ultra-pure source of Uranium-238 (with a radioactivity of about 0.26 nCi and a half-life of 4.51 billion years), allowing the determination of the elapsed time since launch by measuring the amount of daughter elements to remaining U238. The 115 images on the disk were encoded in analog form. The sound selections (including greetings in fifty-five languages, thirty-five sounds, natural and man-made, and portions of twenty-seven musical pieces) are designed for playback at 1,000 rpm, much faster than the 33 rpm speed typical of record players of the era.

Although launched 16 days after *Voyager 2*, *Voyager 1's* trajectory was the quicker one to Jupiter. On 15 December 1977, while both spacecraft were in the asteroid belt, *Voyager 1* surpassed *Voyager 2's* distance from the Sun. *Voyager 1* then proceeded to Jupiter (making its closest approach on 5 March 1979) and Saturn (with closest approach on 12 November 1980). Both prior to and after planetary encounters observations were made of the interplanetary medium. Some 18,000 images of Jupiter and its satellites were taken by *Voyager 1*. In addition, roughly 16,000 images of Saturn, its rings and satellites were obtained.

After its encounter with Saturn, *Voyager 1* remained relatively quiescent, continuing to make in situ observations of the interplanetary environment and UV observations of stars. After nearly 9 years of dormancy, *Voyager 1's* cameras were once again turned on to take a series of pictures.

On 14 February 1990, *Voyager 1* looked back from whence it came and took the first "family portrait" of the Solar System, a mosaic of sixty frames of the Sun and six of the planets (Venus, Earth, Jupiter, Saturn, Uranus, and Neptune) as seen from "outside" the Solar System. After this final look back, the cameras on *Voyager 1* were once again turned off. As noted, the spacecraft continues to provide data transmissions to Earth, signals that currently—even at light speed—require some 17 h.

The Archaeology of Autonomous Machines

Because of their distance from Earth and the resulting time-lag for commands, the Voyager spacecraft were designed to operate in a highly-autonomous manner. Yet by logical extension this brings us into a dicey area of inquiry, if for no more reason than its seemingly obvious implication that tools of human ingenuity are increasingly less our servant and more our master. As Lord Kenneth Clark wrote more than a quarter of a century ago: "[Machines] have ceased to be tools and have begun to give us directions" (Clark 1969: 346). Aerospace exploration increasingly is conducted remotely, marking a turning point in our theoretical considerations of biocultural adaptation.

Explorers of the polar regions of the Earth in the late 19th century took the first steps in removing the burden of expedition transport from the backs of men and elevating the human senses over the Polar Sea when they adopted balloons and dirigibles for their explorations. Now the human senses have been removed from the equation altogether and replaced by the pure intellect of an individual in front of a telemetric monitoring station, and of the potential archaeological record created by the actions of machines that in some cases operate on the other side of the System from the hand and mind that presumably control them.

This has implications for biological anthropologists, but probably not ones they will at first be comfortable in considering at great length. By effectively removing the environment as a calculation in the adaptive value of technology, we have taken the first step in developing a theoretical basis for the study, not of human behavior, but the—for want of a better term—*behavior* of the machines that have begun to explore for us, and how that behavior reflects an originating culture that in future years may exist many light years distant.

These are considerations that go far beyond one individual or one expedition, and lead into general considerations of the courses of technological development and human expansion, and the kinds of imaginative notions that contribute either to success or failure in all explorations of hostile environments.

Conclusions

Once a spacecraft no longer responds to signals from Earth, it ceases to be used for the original mission for which it was designed, and becomes instead a discarded, and hence, archaeological, object. This is the case with the *Pioneer 10* space probe, which ceased 'speaking' with Earth in 2003, and is now headed on a two million year journey toward the red star Aldebaran. In a similar way, the *Voyager 1* probe has recently crossed the boundary the separates the Solar System from interstellar space and now enters a journey of almost infinite proportions, especially in relation to the life span of *Homo sapiens* as individuals and even as a species.

One is left to contemplate the infinite varieties of responses these two different probes—with their two very different approaches to interstellar cultural communications—might receive. There is not only the difference in messages—one almost purely physical and biological and the other almost comprehensively cultural—but the unknown, and ultimately unknowable, variety of extraterrestrial civilizations that might 1 day be in a position to intercept and study these artifacts in the same manner as traditional terrestrial archaeologists.

But in space—as on Earth—the strict categories of systemic and archaeological context are not absolute. It is possible for an object to move in and out of context.

These categories of archaeological methodology are especially tricky in the aerospace environment, but absolutely necessary if we are to consider the possibilities of fieldwork on the now dead—or soon-to-be dead—exploring spacecraft humans have launched into the void.

References

Capelotti, P. J. (2010). *The human archaeology of space: Lunar, planetary and interstellar relics of exploration.* Jefferson, NC: McFarland.

Clark, K. (1969). *Civilizations.* London: BBC Books.

Deetz, J. (1977). *In small things forgotten: The archaeology of early American life.* New York: Doubleday, Anchor Press.

Dymond, D. P. (1974). *Archaeology and history: A plea for reconciliation.* London: Thames and Hudson.

Fimmel, R. O., & Swindell, W. (1977). *Pioneer odyssey.* Washington, DC: NASA.

Finney, B. R. (1992). *From sea to space.* Palmerston North, New Zealand: Massey University Press.

Gould, R. A. (1983). *Shipwreck anthropology.* Albuquerque: University of New Mexico Press.

National Aeronautics and Space Administration (NASA). (1972). *Analysis of Surveyor 3 material and photographs returned by Apollo 12.* Washington, DC: NASA, Scientific and Technical Information Office.

National Aeronautics and Space Administration (NASA). (2013). NASA spacecraft embarks on historic journey into interstellar space, September 13, 2013. Accessed November 29, 2013. http://www.jpl.nasa.gov/news/news.php?release=2013-277.

National Space Science Data Center (NSSDC). Pioneer 10. National Space Data Center. Accessed November 29, 2013. http://nssdc.gsfc.nasa.gov/nmc/masterCatalog.do?sc=1972-012A.

Nickle, N. L. (1971). Surveyor III material analysis program. In *Proceedings of the Second Lunar Science Conference* (Vol. 3, pp. 2683–2697). Boston: MIT Press.

Raistrick, A. (1972). *Industrial archaeology: An historical survey.* York: Methuen Publishing Limited.

Schiffer, M. B. (1987). *Formation processes of the archaeological record.* Salt Lake City: University of Utah Press.

Steinberg, R. (2000). *Dead Tech: A guide to the archaeology of tomorrow.* San Francisco: Sierra Club Books.

Chapter 5
The Space Shuttle *Discovery,* Its Scientific Legacy in a Museum Context

Hanna M. Szczepanowska

Abstract This chapter is written from the perspective of a conservation scientist who works on early satellites, solar panels, thermal protection systems and heat shields from the Apollo and Space Shuttle missions in the context of museum exhibits. The author looks at the Space Shuttle orbiters as technologically advanced systems which, when their missions were completed, entered museums as part of humanity's scientific and cultural heritage.

Introduction

The operation of complex, integrated space systems requires revolutionary thinking both in their development and management. Space exploration, whether human or robotic, is the grandest and most technically challenging expression of human imagination and ingenuity.

Theodore von Karman said: "Scientists study the world as it is; engineers create the world that has never been" (after Griffin 2008). However, analysis of engineering creations and the prediction of their behavior under various environmental and operational conditions require science. The synergy of both produces unprecedented results. Spacecraft, upon ending their operation begin a new phase while on exhibit in the science museums, educating the public about the space exploration. Some of them, reaching their 50 anniversary of creation, form a new collection of space-archeology artifacts. How these artifacts can be presented in a museum setting to fully experience their performance in the space environment and harsh atmospheric re-entry? A proposed concept of showcasing the space heritage artifacts integrates both humanistic and engineering approach.

An erratum to this chapter is available at DOI 10.1007/978-3-319-07866-3_10

H.M. Szczepanowska (✉)
MCI Smithsonian Institution, Suitland, MD 20746, USA
e-mail: hszczepanowska1@gmail.com

© Springer International Publishing Switzerland 2015
B.L. O'Leary and P.J. Capelotti (eds.), *Archaeology and Heritage of the Human Movement into Space*, Space and Society, DOI 10.1007/978-3-319-07866-3_5

The *Discovery,* one of the Space Shuttle Fleet Orbiters, completed its last 133rd mission on March 9, 2011. One month later, it retired to its new permanent location at the National Air and Space Museum, Udvar Hazy Center, in Chantilly VA (NASA 2012). This unique spacecraft that traveled to the Low Earth Orbit (LEO) and returned to Earth embodies engineering achievements measured by the importance of its science missions and design of the spacecraft. To the public, the space shuttle delivered results as the program promised; routine launches of cargo and people into orbit, returning to Earth for refurbishing and launching again for a new mission. Although it seemed routine, the vehicle, its maintenance and the missions all were highly complex.

The focus of this discussion is on the technological legacy exemplified by the thermal protective system (TPS) of the space shuttle orbiter which posed one of the main challenges in ensuring the shuttles' safe return to Earth. These technological advancements are highlighted in the context of TPS predecessors, ablative systems of Apollo, Gemini, and Mercury which by now have gained the status of space archaeology objects. How that technology can be experienced by and conveyed to the museum visitors during their short contact with an artifact on exhibit is proposed in a new framework of visitors' interactions with these artifacts.

The Space Shuttle Program, an Overview

The phrase 'space shuttle' refers to the program of Space Transportation System (STS) which begun its operation in 1981 with the first STS flight of *Columbia,* on April 12. The first flight in 1981 commenced the three decades lasting operation of the space shuttle program. The 133th and final launch was on March 9, 2011, from the Kennedy Space Center.

All five orbiters of the fleet, *Columbia, Challenger, Endeavour, Discovery* and *Atlantis* were flown. A test orbiter, *Enterprise*, was built for the purpose of testing approach and landing and did not have capability to fly into orbit; it lacked engines, heat shields and any equipment required for orbital flights. After completion of tests it became part of the collection at the National Air and Space Museum (NASM) exhibited since 1985 at the new branch of NASM, the Steven F. Udvar-Hazy Center in Chantilly, VA. On April 17 of 2012 it was replaced by *Discovery* once the Space Shuttle Program was no longer in operation (Fig. 5.1).

Similarly, all remaining orbiters continue their educational role in various science museums and centers across the United States. The *Enterprise* was transferred to the Intrepid Sea-Air-Space Museum in New York City shortly after leaving the Udvar-Hazy Center in 2012. *Atlantis* is on display at the Kennedy Space Center Visitor Complex in Florida. The *Endeavour* became part of the exhibit at the California Science Center in Los Angeles, CA.

Two shuttles, *Challenger* and *Columbia,* were lost in missions in 1986 and 2003 respectively. In 1991, the orbiter *Endeavour* was built as replacement of *Challenger.* The Space Shuttle vehicles delivered payloads to orbit, re-entered the atmosphere and captured large payloads on their return back to Earth. STS orbiters were the first space vehicles to travel multiple times to LEO and back to Earth.

Fig. 5.1 *Enterprise* and *Discovery* facing each other on the grounds of NASM's Udvar-Hazy Center, April 17, 2012 (image 2012, H.Szczepanowska 2012)

One of the most notable functions of the STS was participation in the construction and later the servicing of the International Space Station. Other missions included collaboration with the European Space Agency and their Spacelab program, supporting numerous scientific experiments that were carried out in space.

The design of the orbiters, their deployment strategies as well as the multifaceted purpose of the shuttle program evolved after many years of changing objectives, adjustments to congressional budgetary cuts, and the desire to incorporate many functions meeting military, scientific and commercial needs. The complex political climate shaping the shuttle program development is the subject of space historians' study. This chapter is focused on the artifact, the orbiter exhibited in a museum, what it represents in that context, how its technological advancements can relate to its predecessors and how that can be traced in its design and, finally, how to present the artifact and its missions to museum visitors often overwhelmed by the sheer size of an artifact that measures 122×78 ft and weighs 171,000 lb.

The Space Shuttle, General Design Requirements

The main objectives of the space shuttle vehicles—to travel to space on various missions and return to Earth—shaped the spacecraft design. Furthermore, the design had to accommodate military requirements for high capacity payload deployment. The unique design of STS orbiters' side doors facilitated deployment of large satellites such as the Hubble Space Telescope.

To meet the requirements at reentry to the Earth atmospheric a new thermal system had to withstand high temperatures and impacts experienced during reentry. The invention of reusable thermal protection systems, 'heat shields' that would not ablate and could be reused to adequately protect each part of the space vehicle exposed to ranges of fluctuating temperatures was one of the greatest challenges.

Each space shuttle included three main assemblies, an orbiter vehicle (OV), a pair of recoverable solid rocket boosters (SRB), and an expandable external tank (ET) with liquid hydrogen and liquid oxygen. All of these components were stacked together. The shuttle had a two-stage ascent and was lifted by its two SRBs reinforced by three main engines fueled by liquid hydrogen and oxygen from the external tank. Two minutes after liftoff the pair of SRBs was separated by frangible nuts that held them in place until that moment. SRBs fell by parachute into the ocean to be recovered for refurbishing and reuse.

Two orbital maneuvering systems (OMS) engines facilitated both jettisoning and, later, the orbiters' drop out of orbit and re-entry to the atmosphere. The launch was vertical, similar to a conventional rocket launch. Once the space mission was completed the orbiter fired its OMS to re-enter the atmosphere to achieve the necessary hypersonic speed (N.B.: In aerodynamics such speed is associated with Mach 5 or above; hypersonic speeds are greater than the speed of sound).

The other great challenge, next to the development of reusable heat shield, was to design a vehicle with aerodynamic stability at various speeds from subsonic and supersonic at points from atmosphere reentry to controlled gliding on landing. The aerodynamic shape was a compromise between the demands of radically different speeds and air pressures during re-entry, hypersonic flight and subsonic atmospheric flight. Large wings were included to accommodate gliding of the shuttle at the end of its descent.

Such space vehicles conceptually existed two decades prior to the Apollo program. The concept of a reusable winged spacecraft and suborbital bomber was developed in the 1940s although it materialized later (Day 2003). Many attempts of designing a reusable and multifunctional spacecraft followed, some of them even overcame technical challenges of reentry using heat shields that did not ablate such as the X-20 Dyna-Soar, however the project canceled before flight trials could begin (NASA 2008a, b).

Maxim Faget, who oversaw the space shuttle design, was also involved in designing space vehicles of Mercury, Gemini and Apollo, so similarities of the early engineering solutions can be traced in the STS orbiters. The legacy of the engineering solutions from the Apollo era is exemplified in the thermal protective systems.

The Historic Context of Thermal Protective Systems

Atmospheric reentry is the movement of human-made or natural objects as they enter the atmosphere of a planet from outer space, in case of earth, from an altitude above Karman Line (100 km; 62.1 miles) (Donegan 2009: 83–90; see also

Darrin this volume). The Thermal Protective System (TPS) is critical for the successful reentry of the space vehicle to the Earth atmosphere; it shields the cargo and human crew inside the vehicle. The space vehicle reentering the atmosphere experiences surface pressure, convective, catalytic and radiative heat, shear heating, vibration, turbulence, just to name a few simultaneously occurring impacts upon the spacecraft. The heated surface interacts with the gas boundary layer. In that environment the composite materials undergo chemical and physical changes (Dymitrienko 1999).

Re-entry parameters dictated the choice of materials which, in order to sustain the impacts of reentry forces, had to possess slow thermal conductivity and large enthalpy. Different thermal management techniques are applied to different flight vehicles. In case of vehicles reentering the Earth atmosphere two different TPS systems were employed in managing the aerodynamic heat and impact of reentry, ablative and reusable systems.

The first manned space vehicles of Apollo era, Gemini and Mercury, used ablative shields; the space shuttle used reusable systems [N.B. The first Mercury spacecraft used a blunt body design and a heat sink; the later version used blunt body design and an ablative material (Swenson et al. 1966)]. Some of the processes used in the development of TPS systems lead to innovative solutions, such as reinforced carbon–carbon (RCC) and reusable silica-base tiles later used on the space shuttle orbiters.

The first ablators were phenolic resin plastics, modeled into desired shape. Nylon cloth was impregnated with the phenolic resin and underwent pyrolysis, a process utilized in the production of STS's RCC. Some of the materials which were considered in the early development of thermal protection systems included: beryllium oxide, ceramics, oak, wet oak, graphite, plastic laminates, and glass cloth saturated in thermosetting resin. A selection of short fibers, randomly oriented in a 'soup' of resin, molded into desired shape was promising. The refractory fibers used in the early experiments included oxides, mostly silica oxide, to hold the charred resin. Another fiber, graphite, was attractive material because its strength and thermal conductivity increase with increasing temperature (Sutton 1982: 3–11). This extensive research in the 1960s lead to even greater technological advancements of TPS in later years.

Two examples of heat shields from the manned space vehicles were selected to exemplify the different technologies used in the historic thermal protective systems, one from Mercury 7 (flown 1962) and the other from the Gemini Capsule (flown 1966). They provide a historic context for the TPS utilized on the Space Shuttle orbiter, bridging the early developments with new technological advancements. Phenolic resin used in the historic ablators had the highest-temperature capability, and as it cross linked it reduced to a char with some structural integrity, especially if refractory fibers were holding it together (cross linking occurs when resin is heated under pressure). Development of pyrolytic graphite in the process of testing ablatives attracted considerable interest at that time. This form of graphite is made by placing a high-temperature form in an atmosphere of gaseous hydrocarbons. The hot surface pyrolyzed, the gas molecules which impinged

Fig. 5.2 Mercury 7, 1962, cross-section through the ablated upper part of the heat shield, illustrates layers of the fiberglass cloth laminate saturated with phenolic resin (NASM 1968-0263-002; image 2008 H. Szczepanowska)

upon it, left the carbon on the surface. This technology was known then as Reinforced Pyrolyzed Plastic (RPP). RPP was used on the door of the capsules and the navigation equipment. The process was revisited later leading to the fabrication of RCC on the space shuttle orbiters. The Mercury heat shield material was cloth-like, fiberglass-reinforced resin (Fig. 5.2). It exemplifies one of the methods of manufacturing high-performance composites which involved the resin-impregnation of bundled filaments to form a continuous tape or fabric. This material was subsequently plied in layers and then subjected to pressure and heat, which yielded a consolidated composite structure for subsequent assembly or manufacturing steps.

The Gemini heat shield utilized honey-comb structure filled with ablative material (Fig. 5.3). Avco invented a method of structurally fastening a low-density polymer to the substructure of metallic honeycomb. The depth of each cell was considerably greater than its width, so that the ablation material adhered to the honeycomb when exposed to shear which ensured that the ablation material did not fall out (Dolan 1965). The same principle of ablative shield was used on Apollo vehicles and is discussed elsewhere (Szczepanowska and Mathia 2011).

The ablative technologies, historically successful in protecting the cargo upon reentry to the Earth atmosphere, could not be employed on the space shuttle; ablative materials could not be reused. However, as pointed out earlier, some of the technological concepts from early materials' tests were developed further and utilized on the orbiters' TPS.

Fig. 5.3 Gemini, 1966, cross-section of a heat shield showing a fiber-glass honeycomb core filled with an elastomeric ablator, Dow-Corning DC-352 (NASM 1968-0580, image 2008, H. Szczepanowska)

The STS Orbiters' Thermal Protective System

The space shuttle orbiter was the first space vehicle designed to be used multiple times; its thermal protection system had to be reused for 100 missions. A heavy thermal protection system such as ablative heat shields that disintegrated during re-entry would not work. The space shuttle, which is much larger than the early single-use spacecraft, needed a system that was lightweight in addition to being reusable.

Among several TPS systems used on the orbiters, two were selected to exemplify one of the most interesting technological advancements, Reinforced Carbon–Carbon (RCC) and High-temperature Reusable Surface Insulation (HRSI) ceramic tile. Both are discussed in the context of the multitude of TPS materials used on the orbiters.

The space shuttle orbiters were exposed to a range of temperatures in different locations of the spacecraft at the time of re-entry to the atmosphere. The external surface reached extreme temperatures up to 3,000 °F (Jenkins 1993). Exposure to temperature depended directly on the orbiter's position upon re-entering the atmosphere. In most cases the orbiter flew nose first and upside down. Firing the RCS thruster pitched the orbiter into nose-first position with the underside exposed to the most extreme heat. Moving at 17,000 mph (28,000 km/h), all the phenomena occurring at that speed, upon contact with air, produced heat reaching 3,000 °F (1,650 °C).

The structural temperatures were required to stay below 350 °F for re-use purposes. To keep the temperatures at that level (down to 350 °F), the re-usable surface insulation tiles were applied to the areas that experienced the highest temperature, the windward surfaces. Reusable blankets were used primarily on other leeward surfaces. Reinforced carbon–carbon (RCC) was applied on the

Fig. 5.4 Reinforced carbon–carbon (RCC) shield applied on the leading edges and nose of the shuttle orbiter (courtesy of Dr. Nathan Jacobs, NASA Glenn Research Center)

leading edges and the nose cap, where temperatures were greater than 2,300 °F (Fig. 5.4). Furthermore, high-temperature coatings of SiC-based were used on the Space Shuttle Orbiter leading edges protecting the surface up to 3,000 °F (Alvaro and Snapp 2011). In addition to embracing the challenges of engineering reusable materials, the attachments to the various components of the vehicle, tank, insulation or sub-structure of TPS, posed additional technical difficulties. These materials operated at a wide range of temperatures, expanding and contracting at different magnitudes.

Other thermal protective materials of the orbiter included black, high temperature, reusable surface insulation tiles (HRSI), which insulted upper and forward fuselage windows. White Nomex blankets were used on the upper payload bay doors, portions of the upper wing and mid/aft fuselage. White tiles, low-temperature reusable surface insulation tiles (LRSI) were applied on the remaining areas, forward, mid-, and aft fuselage, vertical tail, upper wing and OMS/RCS pods, where temperature did not reach 1,200 °F (NSTS 1988).

Reinforce Carbon–Carbon

RCC development can be traced to the Apollo era, more precisely, to research and development of Reinforced Pyrolyzed Plastic (RPP). The idea of pyrolyzing a polymer to carbon thus led to RCC (12/10/2013, Personal communication with Dr. Nathan Jacobson, Research Physicist, NASA Glenn Research Center, Cleveland, OH who conveyed this information from David Wright at Lockheed). Fabrication of RCC began with pyrolized (or graphitizing) rayon cloth impregnated with phenolic resin. The cloth layers which formed a laminate were cured in an autoclave. The rayon fibers were pyrolyzed to become carbon fibers.

The two-step pyrolizes involved, first, the conversion of resin to carbon and, next, treatment with furfural alcohol. The carbon-cloth was repeatedly infiltrated with a carbon precursor to form a carbon matrix. Each infiltration step was designated RCC-0, RCC-1, RCC-2, etc. That produced the carbon fibers in a carbon matrix, hence the name carbon/carbon, i.e. carbon fibers in a carbon matrix.

To ensure reusability of RCC, the outer layer is converted to silicon-carbide, protecting the surface against oxidation. Once converted, the top layer is whitish-gray color, with cracks caused by differential thermal expansion throughout the curing and oxidation processes. The operating temperature of RCC ranged from −250 to 3,000 °F. The final product, having a relatively high thermal conductivity with respect to other thermal protection system components, promoted the internal cross-radiation from the hot stagnation region at the apex to cooler areas of the component. This cross-radiation reduced the temperatures near the apex (NASA 2008a).

The reinforced carbon–carbon (RCC) was used on the nose cap and an area immediately aft of the nose cap on the underside and on the leading edge of the wings. RCC protected the areas where temperature exceeded 2,300 °F. The leading edges of each of the orbiters' wings had 22 RCC panels. The molded components were approximately 0.25–0.5-in. thick.

During fabrication, the RCC panels were covered with a silicon carbide coating; a final coating of glass served as sealant. Early RCC did not use a glass protection system. The idea of protecting RCC with a glass developed after the first few STS flights and was a major step forward (personal communication with Dr. N. Jacobson, NASA Glenn). Although the RCC panels were strong and capable of withstanding extreme temperatures, they were thermally conductive, which required to use insulating blankets and tiles behind the RCC panels to protect the internal structures of the orbiter.

High Temperature Reusable Surface Insulation Tiles

Several different types of reusable tiles were used on the STS orbiters. Among them were Advanced Flexible Reusable Insulation (AFRSI), Flexible Reusable Surface Insulation (FRSI) and several types of ceramic tiles. The two main categories of ceramic tiles included High-temperature Reusable Surface Insulation (HRSI) and Low-temperature Reusable Surface Insulation (LRSI), differentiated by color coating, pigmented with materials that would respond to different ranges of temperature.

HRSI, found on the lower surface of the orbiter were coated with black borosilicate glass, while LRSI tiles were coated white (Alvaro and Snapp 2011). The white coating contained silica and alumina and better reflected the heat of the Sun while on–orbit. They were used for areas exposed to lower temperatures, ranging from 600 to 1,200 °F. The white tiles were usually larger and thinner, 8 in. long on each side and from less than a 1/2 in. thick up to 1 in. in thickness. Waterproofing polymer was applied as the final coating on all tiles.

HRSI, once made of the ceramic tiles, were of different thickness, depending on the temperature that the surface was exposed to; thicker at the forward areas of the orbiter and thinner toward the aft end. Except for closeout areas, the HRSI tiles were nominally 6- by 6-in. squares.

Black, reusable tile exemplify another technologically advanced material of which conceptual and empirical roots can be traced back to the mid-1950s. From a broad range of material used in the early research from 1957, silica was selected as the primary component by 1961. The final component of the basic shuttle material emerged in 1968 as LI-900 (Lockheed Insulation/9 lbs per cubic foot). It is a low-density and high-strength rigid fiber ceramic tile, meeting the main objectives, to protect against high temperatures, of light weight and reusable.

The reusable surface insulation ceramic tiles were used on the entire fleet of shuttle orbiters. The earlier orbiters used 34,000 tiles, later only 26,000, replacing the areas of exposure to moderate re-entry temperatures with flexible insulting blankets.

High purity amorphous silica fibers derived from sand and were the main component of bulk of these ceramic tiles. The fibers of 2–4 micron in diameter are approximately 1/16th in. long suspended in water slurry were cast, forming soft, porous blocks; colloidal silica binder was added to hold them together. Dried and sintered at 2,300 °F, the blocks were cut to precise dimensions. Each tile was unique to fit the curvature of surface on the orbiter. Machined tiles were covered with coatings, baked-on in ovens. The black coating of borosilicate glass covered the tiles that experienced the highest temperatures at re-entry, up to 23,000 °F (that is why they are referred to as high-temperature tiles). Not surprisingly, after atmospheric re-entry, this coating shows only minor changes of its surface texture and roughness (Szczepanowska and Renegar, 2014).

An uncoated HRSI tile held in the hand feels like very light foam, less dense than Styrofoam, and the delicate, friable material must be handled with extreme care to prevent damage. The ceramic coating, forming a thin, hard shell encapsulates the friable fibers on all sides except the side where the tile is attached to the orbiter's surface (Fig. 5.5). Even a coated tile feels very light, lighter than a same-sized block of Styrofoam.

Exhibit

The STS Orbiters represent the class of space artifacts embodying multitude of technological challenges, not least those encountered in engineering solutions to atmospheric reentry. The thermal protective systems discussed earlier exemplify one aspect only, innovative systems and materials designed for multiple re-entry to the Earth atmosphere, thus serving as a model applicable to analysis of space

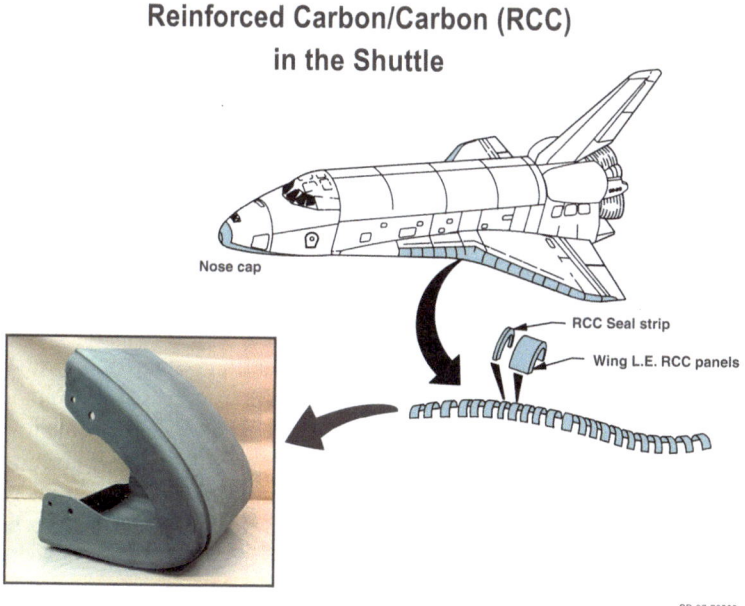

Fig. 5.5 Detail of HRSI tile reveals in the chip areas structure under the black coating, RTV adhesive and remnants of an insulation pad to which it was attached.(Image H. Szczepanowska 2008)

collections in general. The primary consideration for a museum curator is how to exhibit those artifacts so they would effectively enable the museum visitors a glimpse into challenges of space exploration.

The technological-scientific artifacts, to which space vehicles belong, are not merely symbols and icons, presented in a narrative of space historians; they were functional, operational, evolving and dynamic structures. That dynamism must be conveyed to the public in an exciting way, aiming to bring closer the experience of space exploration.

A discussion on this subject—among historians entrusted with care for space artifacts in museums—lead to a meeting entitled "Artefacts" in 1996. One of the objectives was to develop a model for effective use of scientific collections in historical studies (Collins and Millard 2005). Although many essays were published after the "Artefacts" meeting, and thought it was clear that an appreciation of the objects as conveyors of the technology grew over the years, the exhibits did not seem to change very much. They remained static and dense sediment of human agency, culture and technology.

Without doubt, word-based information is essential in conveying concepts of knowledge embodied in artifacts, which that is particularly important when exhibiting space artifacts. Many scientific definitions and concepts originated in specific disciplines, thus the need for clearly defining terms and developing word-base communication is even more essential.

However, to fully appreciate an artifact especially one involved in space exploration, calls for a full-sensory experience, engaging olfactory sensations of charring and melting of heat shields, acoustic impressions of supersonic speed, and the many other imaginable effects of reentry impact. This framework of presenting space artifacts expands the traditional means of exhibits and reaches out into the epistemological concept of inquiring information, through non-verbal means.

Epistemic culture occupied with conveying knowledge distinguishes between settings of generating such knowledge and stressing its contextual aspects. Such a concept is particularly appropriate to experiencing space objects and the science which they represent, in order to understand how knowledge is created and how it can be perceived through these scientific objects.

Cultural values and ideas, a subject of historical and anthropological studies, serve as a word-base background for tailoring definitions necessary to communicate scientific terminology and concepts. Furthermore, it is nearly inescapable to include the political background of space exploration, which provides another layer of word-based interpretation for artifacts from that era. That, enriched with an understanding of the technology which enabled the objects to function, formulates a base for designing a full-sensory experience for experiencing the space objects on exhibit.

Conclusions

Space artifacts exemplify particularly rich, intriguing, and fascinating, multi-layered complex technological systems. The tracing of the technological innovations encapsulated in an aerospace artifact was illustrated utilizing examples of thermal protective systems to elucidate some aspects of complexities involved in designing space vehicles and innovative materials suited for space missions.

Details of TPS technology developed for STS orbiters that traveled multiple times through the earth atmosphere, characterization of some of many materials used in TPS and analysis of their behavior in varied environments, provide examples of the nearly limitless possibilities that accompany the presentation of a complex artifact in a museum setting. The core of this model is to streamline the concept of complex technology embodied in themes and subjects represented by space artifacts.

The space shuttle orbiters exemplify multitude of technological advancements of engineering and material science. The multifaceted objectives of their mission and the international collaborations provide a full spectrum of possibilities encapsulated in the artifact itself should inspire and engage the exhibition of space technologies. Today's technology offers a multitude of interactive solutions, which, when coupled with the scientific approach of monitoring the receptions and responses of museum visitors, would provide a platform for quantitative and qualitative analysis of the success of such exhibits. This new framework should invigorate and bring much closer some aspects of the space exploration adventure while educating the audience about complex technological systems.

References

Alvaro, R. C., & Snapp, C. G. (2011). *Orbiter thermal protective system lesson learned.* AIAA 2011-7308, AIAA Space 2011 Conference and Exposition, Long Beach, CA.

Collins, M., & Millard, D. (Eds.). (2005). *Showcasing space.* Cromwell Press, NMSI Trading Ltd, Science Museum.

Day, A. D. (2003). *Early reentry vehicles: Blunt bodies and ablatives.* US Centennial of Flight Commission, 1903–2003. http://www.centennialofflight.net/essay/Evolution_of_Technology/reentry/Tech19.htm

Dolan, C. M. (1965). *Study for development of elastomeric thermal shield materials.* Prepared under contract No. NAS 1-3251 by General Electric Company, Philadelphia PA for NASA, March 1965. NASA CR-186.

Donegan, M. (2009). Space basics: Getting to and staying in space. In A. G. Darrin & B. L. O'Leary (Eds.), *Handbook of space engineering, archaeology and heritage.* Boca Raton: CRC Press, Taylor and Francis Group.

Dymitrienko, Y. I. (1999). *Thermodynamics of composites under high temperatures. Series: Solid mechanics and its applications* (Vol. 65). Dordrecht: Kluwer Academic Publisher.

Griffin, M. (2008). *Leadership in space: Selected speeches of NASA administrator Michael Griffin,* May 2005–October 2008. NASA SP-2008-564.

Jenkins, D. J. (1993). *The history of developing the national space transportation system: The beginning through STS-50.* Marceline, MO: Walsworth Publishing Company.

NASA. (2012). Media Advisory. M12-062, April 9, 2012. Accessed January 3, 2014. http://www.nasa.gov/home/hqnews/2012/apr/HQ_M12-_SCA_Discovery_Flight_DC.html

NASA. (2008a). *NASA: The first 50 years.* An aerospace America special report. Washington, DC: NASA.

NASA. (2008b). Orbiter thermal protection systems. NASA Facts-FS-2008-02-042-KSC. FL: Kennedy Space Center. Accessed January 3, 2014. http://www.nasa.gov/centers/kennedy/pdf/167473main_TPS-08.pdf

NSTS. (1988). *NSTS 1988 news reference manual.* Accessed January 3, 2014. http://science.ksc.nasa.gov/shuttle/technology/sts-newsref/sts_sys.html#sts-rcc

Sutton, G. W. (1982). *The initial development of ablation heat protection: An historic perspective.* Everett, MA.: Avco-Everett Research Laboratory, Inc. (AIAA 50th Anniversary 1981: *19*(1): 3–11).

Swenson, L. S., Jr., Grimwood, J. M., & Alexander, C. C. (1966). *This new ocean: A history of project mercury.* NASA SP-4201, Washington, DC.

Szczepanowska, H., & Mathia, T. G. (2011). Space heritage: The apollo heat shield; atmospheric reentry imprint on materials' surface. *Materials Research Society Symposium Proceedings, 1319.* doi:10.1557/opl.2011.780.

Szczepanowska, H., & Renegar, Th. (2014, in press). Space Exploration Heritage; Characterization of HRSI Ceramic Titles in the Space Shuttle Program using Surface Metrology Techniques. MS&T 14, October 2014, Pittsburgh PA. *Materials Science & Technology 2014, Collected Proceedings.*

Chapter 6
Purposeful Ephemera: The Implications of Self-Destructing Space Technology for the Future Practice of Archaeology

Justin St. P. Walsh

Abstract This chapter is presented from the perspective of a professional archaeologist who specializes in Greek archaeology, intercultural contact and exchange, and the ethics of cultural heritage. His chapter investigates the mandates for discard and "design for demise" of space objects in the wider context of cultural phenomena from all cultures. The chapter finds comparanda for purposeful ephemera in examples from the media of performance, architecture, and visual art.

Introduction

Recent trends in space mission design are likely to have serious consequences for future archaeological research on the development of technology. These trends, especially the introduction of new international standards governing the lifespans of equipment launched into Earth orbit, have resulted from the consequences of past and ongoing space missions as well as the foreseen impact of future activity by national, academic, and commercial participants in spaceflight. The new limits imposed on orbiting objects have been interpreted by the spacefaring community as requiring the destruction of those objects, to the extent (it is hoped) that no material trace of them will remain.

In the strictest sense, this paper is about problems that future archaeologists will face when they attempt to reconstruct the development of technology for the exploration and exploitation of space without direct access to examples of that technology. These problems are analogous to larger transformations in the contemporary world, however, as the objects we live with become increasingly (and purposely) ephemeral in the face of environmentalist efforts to promote recycling and corporate industrial design that favors "planned obsolescence" over long-term

J.St. P. Walsh (✉)
Chapman University, Orange, CA, USA
e-mail: jstpwalsh@chapman.edu

© Springer International Publishing Switzerland 2015
B.L. O'Leary and P.J. Capelotti (eds.), *Archaeology and Heritage of the Human Movement into Space*, Space and Society, DOI 10.1007/978-3-319-07866-3_6

durability. The focus of archaeological observations on tangible, material remains of past human activities has precluded much consideration of ephemeral objects, but it is clear that new methods will have to be developed to study cultural goods that leave no direct evidence of their existence.

The Problem of Space Debris

The definition of new space technology as ephemera was raised in April 2012, at a conference concerning "end-of-mission disposal and requirements" held by the American Institute of Aeronautics and Astronautics at the Jet Propulsion Laboratory. The attendees of this conference, all professionals with long-standing experience in the military and commercial space industries, were previously unaware of efforts by archaeologists to preserve in situ space heritage and of proposals that heritage management should be mandated for missions which are likely to be historic in nature (e.g., Darrin and O'Leary 2009; Capelotti 2010; Walsh 2012). The concerns of these professionals were governed instead by international treaties, national laws and policies, economic resources, mission requirements, and the legacy of past practices.

This last factor seemed, in fact, to dominate discussion, particularly as it was connected to the current problem of space debris. According to the US Air Force's Joint Space Operations Center, which monitors orbital space, as of 15 March 2013, there were roughly 22,000 objects larger than 10 cm in orbit; of these, only 5 % are functioning payloads or satellites, 8 % are rocket bodies or parts, and 87 % are debris or inactive satellites (Vandenburg 2012).

In fact, the actual number of objects in space is much larger, "if pieces smaller than ten centimeters are included, ranging from 500,000 into the many millions" (Chodas 2002). For example, in 1963, the US Air Force dispersed 480 million copper needles, each approximately 2 cm in length, at an orbit altitude of roughly 3,500–3,800 km (2,174–2,361 miles; by contrast, only 2 years earlier, there had only been 54 human-made objects in space) (Overhage and Radford 1964). This launch was part of Project West Ford, which hoped to create a kind of reflective antenna for terrestrial radio signals in the days before widespread deployment of satellite communications (Wiedemann et al. 2001). Despite an initial prediction that the needles would have an orbital lifetime of 3–6 years, at least 46 clumps of these needles were still in orbit more than 50 years later.

The danger presented by orbital debris is significant. Many pieces are fragments, of widely varying sizes. One object, designated J002E3 upon its discovery in 2002, measured 10 m in length and was probably one stage of a Saturn V rocket (perhaps from Apollo 12 in November 1969 (Chodas 2002)). Worse, the problem is growing due to events such as the Chinese test of an anti-satellite missile in January 2007, which led to the creation of over 2,000 trackable fragments and perhaps another 150,000 smaller pieces. A communications satellite, Iridium 33, was destroyed during a collision with a defunct Soviet military satellite, Kosmos-2251,

Fig. 6.1 A crack in the windshield of the Space Shuttle orbiter *Challenger*. The crack was created by collision with a fleck of paint during STS-7 (18–24 June 1983) (NASA)

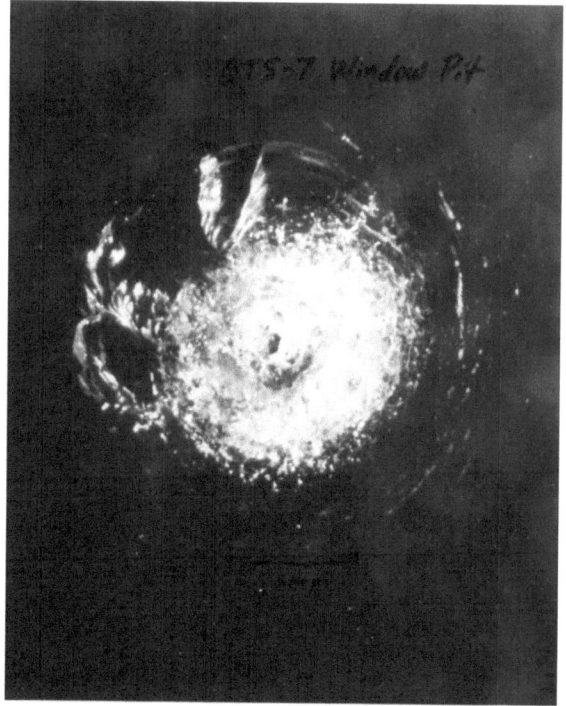

in February 2009, creating yet more debris. A vice-president for Iridium was later quoted as saying that the company receives 400 close approach warnings every week for its fleet of 66 satellites (Weeden blog post).

The crew of the International Space Station has been forced to take cover on several occasions, on account of close encounters with other objects (Drew 2012). More worryingly, in 1983, the shuttle *Challenger* (STS-7) was struck in the windshield by a 0.2 mm fleck of paint traveling at roughly 28,000 kph (Fig. 6.1); the craft could have been depressurized as a result (Hyde et al. 2001: 191–196).

Defunct objects, of course, also pose a potential threat to life and property on the Earth's surface if they de-orbit. Kosmos-954 de-orbited unexpectedly in 1978, spreading radioactive material from its power supply across a 600-km stretch of north-central Canada, and NASA's Skylab showered southwestern Australia with debris in 1979. To date, only one person has actually been hit by deorbiting debris: a woman in Tulsa, Oklahoma, was struck by a piece of a Delta II rocket that re-entered the atmosphere in 1997, 1 year after its launch (Fig. 6.2).

At the same time that debris has been increasing, several developments are opening space up to a wide variety of new participants. For example, the Google Lunar X Prize is encouraging private groups to design a lunar rover and send it to the Moon. Some of these groups, such as Moon Express, see the competition as a stepping stone to exploitation of space resources, which will require much greater traffic.

Fig. 6.2 Lottie Williams of
Tulsa, Oklahoma, with the
piece of a Delta II rocket
which struck her upon
re-entry in January 1997
(courtesy of *Tulsa World*,
used by permission)

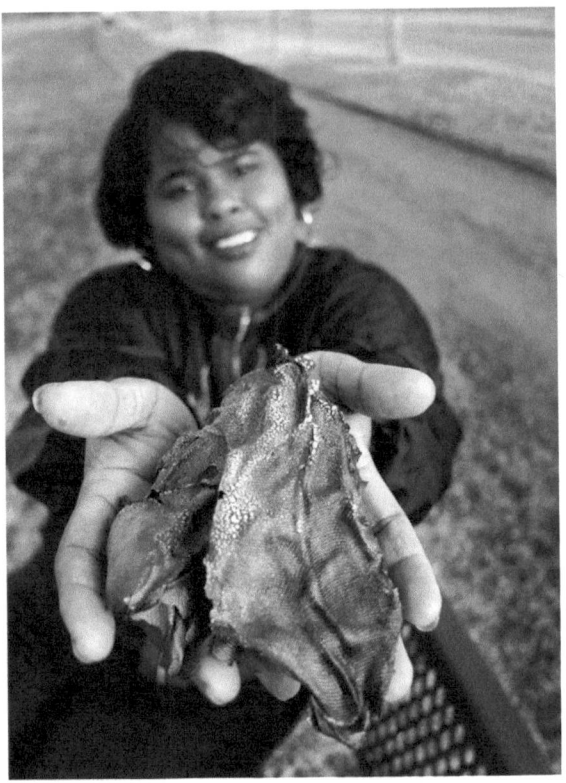

The spacefaring community has already increased its activity, especially in low-Earth orbit, with the development of nano-satellites, or "cube-sats" (Fig. 6.3). This class of equipment is designed to fit inside a standard metal cube which measures 10 cm on a side (or which is made of several such modules linked together). The miniaturization of electronics has made these tiny satellites feasible to construct and useful for research, while the lower cost of producing and launching them compared to traditional satellites is a strong incentive for their adoption by scientists. In one recent launch, NASA placed 29 such satellites into space simultaneously.

One direct consequence of this problem has been the development of new satellites which will themselves clear low-Earth orbital altitudes of debris. These satellites are being developed by scientists in Japan, Switzerland, the US, and likely elsewhere, too; they are still in the testing phase, but their developers hope to deploy them within the next few years (Kawamoto et al. n.d.; Gass and Grosse 2012; Chang 2012). These satellites are intended to "sweep up" debris, collecting it and destroying it by de-orbiting into the atmosphere with it.

Another consequence has been the institution of a new international standard by the Inter-Agency Space Debris Coordination Committee, which is made up of all of the major national and international space agencies. Its Mitigation Working Group issued guidelines in September 2007 (also accepted by the UN through Committee on Peaceful Uses of Outer Space):

Fig. 6.3 Recent
developments in the
miniaturization of electronics
for communication have
allowed satellites to become
smaller and cheaper. One
example of this phenomenon
is the nano-satellite, or "cube-
sat," which puts all of the
hardware within a standard
10 cm cube (or multiple
connected modules of this
size), as shown here (NASA)

A spacecraft or orbital stage should be left in an orbit in which, using an accepted nominal projection for solar activity, atmospheric drag will limit the orbital lifetime after completion of operations. A study on the effect of post-mission orbital lifetime limitation on collision rate and debris population growth has been performed by the IADC. This IADC and some other studies and a number of existing national guidelines [sic] have found 25 years to be a reasonable and appropriate lifetime limit. If a spacecraft or orbital stage is to be disposed of by re-entry into the atmosphere, debris that survives to reach the surface of the Earth should not pose an undue risk to people or property. This may be accomplished by limiting the amount of surviving debris or confining the debris to uninhabited regions, such as broad ocean areas (Inter-Agency Space Debris Coordination Committee 2007).

The IADC document was preceded in March 2007 by a UN Committee on the Peaceful Uses of Outer Space (COPUOS) policy which provided less-defined guidelines, specifically Guideline 6 in the Annex to the Report on the Sixty-Second Session of the General Assembly (United Nations 2007: 50):

Limit the long-term presence of spacecraft and launch vehicle orbital stages in the low-Earth orbit (LEO) region after the end of their mission: Spacecraft and launch vehicle orbital stages that have terminated their operational phases in orbits that pass through the LEO region should be removed from orbit in a controlled fashion. If this is not possible, they should be disposed of in orbits that avoid their long-term presence in the LEO region.

Although this standard is not widely known outside the space industry, it has now also been codified by the International Organization for Standardization for low-Earth orbit (LEO, below an altitude of 2,000 km (1,242 miles), the area where most satellites have been placed) in a document put forth in February 2011 (ISO 2011). Section 6.1.1.2 of the ISO document reads: "Space debris released into Earth orbit as part of normal operations ... shall remain outside the [geostationary orbit] protected region and limit their presence in the LEO protected region to a maximum of 25 years after their release." Anecdotal evidence indicates that the new standard appears to be broadly accepted by commercial industries for future missions.

Criticism can easily be made of the standard's anthropocentrism, given that it suggests that debris including toxic chemicals and radioactive substances should be directed into areas that are uninhabited by humans but are full of other forms of life, such as the world's oceans. In any event, a few scientists have begun to take up the challenges posed by these policies to mission design. For example, Aerospace Corporation's Center for Orbital and Reentry Debris Studies is working on a project based on the principle of "design for demise": creating space equipment from materials specifically chosen for their ability to burn up entirely in the atmosphere during de-orbit re-entry, so that nothing remains to fall to the Earth's surface (see the contributions in Ailor and Wilde 2013: 603–776). The group launches satellites made of different kinds of materials, noting which ones dissipate and which remain following descent. Future satellites will be designed to take advantage of this research and protect the public.

Current events in space mission design therefore pose a problem for future archaeologists who want to study the development of human technology. For, although the study of artifacts is not the only method available to archaeologists or used by them, direct observation of artifacts has long been at the core of archaeological practice—and would be the best way to learn about how the artifacts themselves changed over time. How will we study the development of a society's technology when the evidence we would most like to have is not just largely vanished, but entirely absent? These problems seem similar to the ones faced by future archaeologists who will trace the development of bottles, cans, and other containers which have been subject to recycling since the 1970s, but the ubiquity of those objects seems to ensure that many will be discarded as garbage rather than recycled. Some direct traces of their existence will almost certainly survive.

The space technology considered here, by contrast, generally consists of unique objects (or very limited numbers of types or series of objects) designed for highly specific purposes. If these artifacts are designed to self-destruct completely, how will they later be identified or understood by future researchers as tools or as cultural objects? On one hand, future archaeologists would probably be able to read some significance into the lacunae. They may be able see that first, around the middle of the twentieth century, humans developed space satellites. Perhaps *Vanguard 1*, the oldest piece of manmade material in space (launched in March 1958), will remain in orbit. But in studies of later periods, even though it will likely be known that people continued to engage with space, no satellites dating to the period 60 or 70 years after the earliest examples will be found. We don't know what future techniques might be developed to deal with these problems, or even what kinds of questions future archaeologists might be tempted to ask; any guesses quickly move us into the realm of science fiction. For example, one technique might include some form of sampling of the upper atmosphere and low-Earth orbital altitudes for chemical residues of the incinerated equipment, much as present-day archaeologists sample soils for evidence of human activities (or the way that cosmologists survey cosmic background radiation for evidence of the Big Bang).

Space Technology as Ephemera

The study of purposeful ephemera has been quite limited in archaeology. This fact is perhaps surprising, given that the archaeological record is only a partial sample of the set of materials which once existed—in other words, material culture is, almost by definition, ephemeral. But most material culture is not *purposely* ephemeral. While humans generally have not expected most of the tools or structures that they create to last forever (at least prior to the introduction of the concept of "planned obsolescence" to modern industrial design (Slade 2006)), they have produced many, if not most, of their creations with an eye towards their durability.

The ephemerality found in the archaeological record is largely a result of the effects of processes over time such as the "natural and cultural transforms" elucidated by Schiffer (e.g., 1972: 156–165, 1987). Therefore it can usually be seen as accidental, or at least unrelated to the design of objects, rather than intentional, and, indeed, some material *does* survive, giving us the record of material culture that forms the basis of much of archaeological research.

Human-made objects survive even in space—as the space debris problem described above suggests, some objects survive longer than their designers might want or in ways that are unexpected. Some objects on the Moon, such as the nylon American flags, will have disintegrated following more than 40 years of exposure to alternating extremes of heat and cold, not to mention radiation. NASA's online *Apollo Lunar Surface Journal* collects a variety of opinions on the present state of the flags, as well as remote sensing imagery that shows whether shadows are cast by flags at the various landing sites (Fincannon 2012).

According to astronaut Charles Duke, who took part in the Apollo 16 mission of 1972, the photographic portrait of his family which he left behind at the landing site showed signs of damage almost immediately from the heat of a lunar day (Jacobs 2009; Fig. 6.4). But these objects were not intended from their very invention to be destroyed, and, in any case, archaeologists are used to working with objects on Earth that have broken or partially disintegrated.

Until our present era, though, it seems that there have been relatively few instances of humans creating objects which will be destroyed on purpose or even self-destruct. This is precisely the plan for orbital space equipment. The probability of any one example's survival, while not impossible, thus becomes extremely unlikely.

Consideration of other historical examples of objects that have been purposely constructed to be ephemeral or to destroy themselves seems to be a worthwhile means of trying to understand how ephemera can be studied and what lessons can

Fig. 6.4 Astronauts left some objects on the Moon not directly related to their missions that form part of the material record of their activity there. This image shows a portrait of the family of Charles Duke (Apollo 16, 21–24 April 1972) in a small plastic bag which he placed on the surface. Duke noted that the photograph started to turn brown almost immediately in the heat of the lunar day (NASA)

be learned about what evidence might be the most useful for future scholars. In brief, it will become clear that two primary possibilities exist for preserving useful evidence: replicating the objects, and recording and documenting ephemeral objects and the activities associated with their creation and use (a third possibility, study of the sites where ephemeral space equipment was developed, produced, and launched, also exists, but it presents less direct access to the actual space equipment; see Donaldson, this volume).

The following discussion will explore these possibilities with a view towards the establishment of priorities. Would it be more useful to create and preserve replicas of ephemeral technology, or to emphasize their thorough documentation? In some ways, these approaches are intertwined and overlap, since recordings and documents can help to vet the accuracy of copies. There are, as will be seen, also significant problems with both approaches.

Documentary Evidence

Some of the greatest challenges to documenting and analyzing ephemera can be found in the world of performance—particularly the history of theater, dance, and associated arts. In drama, there has long been a widely-recognized tension between the importance of the authorial intention and the text, and the interpretation of the text by actors, directors, and stage crew—the reconstruction of performances, especially in the far-distant past is highly problematic.

For example, the only known example of stage notes from the sixteenth or seventeenth centuries (Dulwich MSS 1, Article 138, folio 8r) records the lines for the lead character in the play *Orlando Furioso*, probably by Robert Greene, who died in 1592 (Greg 1922; an image of the part can be viewed online at http://www.henslowe-alleyn.org.uk/images/MSS-1/Article-138/08r.html). The notes have been identified as probably by or for the actor Edward Alleyne in a production of 1591 or 1592. It is clear from what remains of this document that there are variations, some significant, between the notes and the same passage in the published first quarto edition of 1594 (Foakes 2005–2013). Lines were added or removed, and their order is altered.

These differences reflect the gap between an artist's conception and the reality of performance, where the same actor will deliver lines differently on different nights, different actors will play the same role differently, and of course, an actor fumbling for a line might even invent new dialogue or skip material in the text. Likewise, the text could be seen as a record only of one specific performance, as opposed to a guide for all performances. In other words, it is important to be wary of treating documents as "ur-texts" encoded with definitive truth.

Contemporary art can draw our attention to similar challenges associated with documenting ephemera. The artist Tino Sehgal calls his art "constructed situations." They are not performance art (according to Sehgal) because they require audience participation and they happen in museums, not theaters. In his pieces, which have taken place in major venues including the Solomon R. Guggenheim Museum in New York and the Tate Modern in London, visitors interact with performers who ask scripted questions or make statements (called "prompts") and converse with them. There are no objects associated with the artworks except costumes and accessories (like umbrellas) held by the participants.

Moreover, Seghal "forbids the creation of any of the by-products—photographs, videos, catalogues, wall text—that normally derive from a work. His pieces leave no physical residue.... He believes that mementos of his work would threaten its purity, which could weaken its effect" (Collins 2012). Even the contracts for his work are oral, rather than written (there are not even any notes taken; later disputes over the contracts are understood to be resolved through the development of consensus among those who were present at the negotiations about what they remembered having occurred). Each time a piece is performed, it will necessarily be different.

The only recordings which exist to document Seghal's works are "tertiary reports," which is to say critical published reviews in newspapers and journals, as well as descriptions such as blog posts by other visitors, and photographs which are illicitly made (visitors are forbidden from making photographs, but in practice it has been impossible to stop them from doing so with cameras on cell-phones; Collins 2012). The lack of detailed documentation makes it practically impossible to study the individual works. Researchers' attention must be directed only towards Seghal's methods instead, but even these are only accessible through the documentation of the few interviews in which he has participated.

Replication as Evidence

One might think that copies of purposeful ephemera would be the most direct way to understand the missing original, but there are complications here as well. The most important Shinto shrine in Japan—the one associated with the *kami*, or spirits, of the Imperial Household, which is located at Ise in southern Honshu—is an especially interesting example of ephemera that has been replicated.

There are two shrines, an inner one and an outer one, at the site. These were first built, according to tradition, in 672 CE (Isozaki 2006). The buildings associated with the two shrines were constructed in a simple fashion from cypress wood and *miscanthus* grass, with copper and gold fittings. The plant-based materials decompose relatively quickly; as a result, a system called *shikinen-sengū* (or *shikinen-zōkan*) has emerged by which, every 20 years, the shrines are totally rebuilt using new materials. 2013 marks the completion of the sixty-second rebuilding.

All of the items dedicated by the imperial family over the centuries (hundreds of sacred treasures and vestments, spinning and weaving tools, military arms and wear, horse equipment, musical instruments, writing implements and daily goods) are also replaced with new versions, the gods are moved in a ritual known as *sengyo*, and the old buildings are dismantled. The former site of the each building becomes a vacant lot which awaits its rebuilding in 20 years.

The *shikinen-sengū* process is thought to preserve traditional Japanese building and religious practices (not to mention metalsmithing, weaving, etc.); in fact, the process of rebuilding became properly systematized by the tradition of cyclical replacement at Ise. The old materials get recycled, either into objects associated with veneration, or into building materials such as metal fittings which are then used at other shrines around Japan. In this case, the physical evidence which remains of the building's shape and function is only what purports to be a perfect, repeated reproduction of the original.

The word "purports" is important here, because it is clear from analysis of the documentary evidence (both photographs and written accounts) that the process has been revised throughout the centuries and not been carried out in the same way every time. It is generally difficult to assess the fidelity of the new structures at Ise to the old ones, since the public is not allowed to enter them, or even to approach them closely. Concentric fences block access and even close views of the shrines.

These fences, however, have themselves increased in number since the nineteenth century, from two to four. Photographs from the earlier period also indicate that the fences were not as tall as they are today (Isozaki 2006: 134). Moreover, the shape and orientation of the buildings has changed, and some Japanese scholars have reconstructed yet more differences going back to the sixteenth century, with changes tending towards increased decoration and complexity as time went on.

More profound problems exist for understanding the original shapes and formats of the Ise shrine buildings, too. In particular, there was a 123-year period (1462–1585) in which no rebuildings happened due to national unrest, and for the last 85 years of that era, there was no shrine at all because "the main precinct" collapsed in 1500 (more recently, the disruption caused by the Second World War caused a delay in rebuilding at Ise, with the date pushed back from 1947 to 1953, followed by a resumption of the traditional twenty-year period; for the information here and below, Fukuyama 1968: 41, quoted and translated in Isozaki 2006: 137). Hardly anyone still alive in 1585 could have remembered the shrine's design and decoration. The response of shrine officials, according to the shrine's own record of the 1586 rebuilding, was to collect information concerning detailed measurements of each structure and [give] it to the carpenters; when this data contradicted the master carpenter's records, officials and master carpenters settled on a compromise. They were forced to determine scale, structure, and form of the new main sanctuary by combining bits and pieces from old records and plans.

As Isozaki (2006) quickly noted, the plans in question could not have been precise even if the shrine had a simpler layout (as seems likely), given the documentation standards of the time. In fact, he suggested, based on the spirit of compromise and "guesswork" recorded in the shrine's document concerning the rebuilding, "a certain will to readjust the design toward a perceived authentic form ... must constantly have been at work" (Isozaki 2006:138). Such a sentiment is a useful reminder of the bias inherent in archaeological reconstruction, too, when investigators make interpretations without extant direct evidence.

These interpretations are supported by (perhaps educated) intuition but also derived from expectations which are themselves generated by previous work, experience with similar situations, and personal attitudes. In any case, a constant process of replication without access to the original, or to the people involved in the creation of the original, is likely to problematic, as it will inevitably introduce variations that are dependent on the interests and personalities of the replicators.

The problems faced by conservators dealing with modern art show the value of recreations, though, even those which happen at a later date. The artist Eve Hesse's 1969 work *Expanded Expansion*, today in the collection of the Guggenheim Museum in New York, was made of reinforced fiberglass poles and rubberized cheesecloth, but it "has become brown and brittle" over time (McCoy 2012). Art conservator Tom Learner described an experimental exhibition during the symposium: "We showed sections of the original piece alongside a material mock-up of a segment that was made by Doug Johns, Hesse's assistant.... Walking past the mock-up was really extraordinary because the slightest air movement caused the latex-impregnated cheese cloth to sway and move—which most definitely did not occur with the embrittled original. You could also smell the materials in the air." He continued, "We're not trying to say that what we've done is the 'right way.' We're trying to explain the different approaches that could be taken, and—ultimately—ensure that the full variety of approaches still remain available to the artwork" (McCoy 2012).

Combinations of Recording and Replication

So far in this discussion, media have been treated uncritically as a means of documenting ephemera—through photographs, drawings, and written descriptions—but media themselves are cultural creations, and they are also subject to the same kinds of ephemerality as other artifacts. Digital media have recently attracted the most attention for the problems associated with its ephemerality, since there are clear problems related to long-term storage, the need for legacy hardware for reading old digital media, and the viability of file formats. These issues direct our attention to the general difficulties associated with preserving any unstable (i.e., cutting-edge, rapidly developing) technology. Some historians (amateur as well as professional) and institutions have been collecting evidence for early software, including software languages (see, for example, the Museum for Computer History, or Paul McJones' *Dusty Decks* blog; http://www.mcjones.org/dustydecks).

The original website created by Tim Berners-Lee at CERN was recreated in 2013 to celebrate the twentieth anniversary of the World Wide Web (http://info.cern.ch/hypertext/WWW/TheProject.html). Many of the links beyond the first page are broken, however—a phenomenon with which most users of the Internet will be familiar when dealing with older, neglected, or incomplete websites. Other projects, such as 404 *page found* (http://www.404pagefound.com/), whose name refers the code used on the Web for an error related to internet resources that cannot be found, attempt to recognize or preserve old websites which still function. The Internet Archive's so-called "Wayback Machine" (http://archive.org/web/) claims to provide access to 368 million old webpages. The Internet Archive was founded in 1996, specifically to provide "a mechanism and a memory" to allow web culture to learn from its successes and failures. It has affiliated with the Smithsonian Institution and other research organizations.

More recently, a remarkable project has begun to address the problem of reconstructing ephemeral media: sounds in the form of recorded audio. Although these media do not fall into the category of purposeful ephemera because they were intended to be durable, their intangible nature provides important insights to help us understand how it might be possible to work with other ephemera. Feaster (2012a) has examined the different means used by humans to record sounds, from manuscripts of musical notation to experimental scientific formats developed in the nineteenth century, and he has developed methods for playing some of these recordings.

The most relevant examples of Feaster's work are probably the recordings made by Emile Berliner, which were pressed onto gramophone records by 1889 and made available for sale beginning in 1890. Several of these records are known only from images of the discs which were printed in contemporary German magazines. Feaster took advantage of the very high image quality used in the printing, which reveals the shape of the recordings' grooves, where the sound was encoded (the text which accompanied the image of the record in the magazine actually gave instructions to readers about how the recording could be recreated from the image, and then played using a homemade bamboo stylus; see Feaster 2012a: 65–66). In one example, the record was a recording of Berliner reading "Der Handschuh" ("The Glove"), a poem written in 1797 by Friedrich Schiller (Fig. 6.5).

As Feaster has described in a blog post (2012b, with playable audio), it was possible to use a computer to play back the sounds by scanning the image at high-resolution, converting the spiral grooves to straight lines, stitching them together, and converting them to a .WAV-format digital audio file by means of a program normally used for digitizing optical film soundtracks. The sounds of the reader's

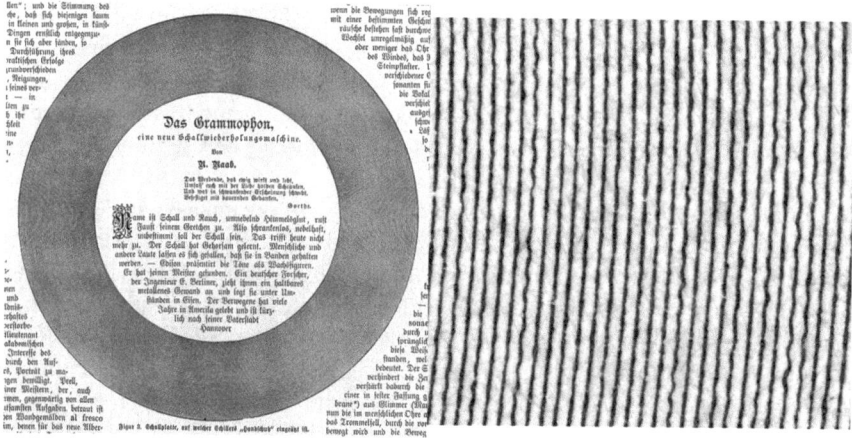

Fig. 6.5 *Left* image of an early gramophone record containing a recording of "Der Handschuh" by Emile Berliner, from the February 1890 issue of the German magazine *Über Land und Meer*. A close-up of the image *right*, showing the shape of the grooves on the record (courtesy of Patrick Feaster)

voice in the resultant file are audible, if not clear. Without a copy of the text available for comparison, it is admittedly very difficult to understand the poem. Even so, the reconstruction of a recording of any fidelity without access to the original matrix on which the recording was made is a remarkable achievement, arguably comparable to producing an accurate recreation of a painting's colors, composition, and textures from a literary description alone. Even better, with the development and distribution of high-quality 3-D printing technology, it is becoming possible for individuals to produce their own playable analog records from raw materials (Ghassaei 2012, with video demonstration), so the possibility exists for a complete recreation of a usable piece of media in the correct physical format without ever having examined the original object.

Conclusion

There are obviously problems associated with what can be learned about original ephemera in each of these cases. Our access is indirect, coming either through visual and literary description, or through the vagaries of copying (which can never be completely faithful) and the destruction and dispersal of the original. Both kinds of access will themselves probably require thoughtful attention to preservation in order to survive for study.

Documentation can describe or depict the original object; it can come in tangible forms (e.g., books or other printed material, photographs or drawings on paper) or intangible (i.e., digital). The examples described above, however, show that documents must be taken as partial and biased, created according to the intentions of the person or people who made the documentation, not with the interests of researchers in mind. Whatever documentation is made and preserved should capture the widest possible range of information relating to the ephemera. The documentation should also be produced in formats which are most likely to be usable in the future, a point which suggests that digital methods, while appealing for their ability to make infinite identical copies, should be used only sparingly until such technology stabilizes.

Copies seem on their face to be more useful than documents, since they can open up new avenues of personal experience with the original (as seen with the recreation of *Expanded Expansion*), but there can still be significant problems. Those that were created closer to the moment of the original's creation are likely more faithful than those created at a later date, or which are at a remove of several object-generations from the original. Just like documents, copies should be treated warily for the biases associated with their creation. Each approach has its own positive qualities, but the problems associated with using documentation or copies show that a combination of these approaches is preferable, and researchers should be aware that documents and copies might be used in innovative ways to gain access to the past (such as 125-year-old sounds).

We already lack access to many original objects related to space exploration, either because they have already been destroyed, or because they remain in remote contexts. For those that still exist, the hope remains that we will someday be able to study them both for themselves and as evidence for humanity's development. The importance of ensuring that such materials as exist in standardized formats likely to be readable by future interested researchers should also be recognized.

The spacefaring community should also be encouraged to preserve experimental versions, prototypes, or replicas of their equipment. Making contact with space scientists and working together with them on these archives thus should be a high priority. Finally, to the extent that a mission can be foreseen to be historic, we ought to promote the preservation of original equipment by boosting it to otherwise-unused higher orbital altitudes.

Acknowledgments I am grateful to Dr. William Ailor of Aerospace Corporation for his invitation to take part in the AIAA panel in March 2012, to Professor Michael Schiffer for his suggestions regarding sampling of the atmosphere, to Professor Stefan Tanaka for directing me to the publications of Arata Isozaki on the Ise shrine, to Mr. Paul McJones for information about the preservation of old software, and to Mr. Patrick Feaster for information regarding the "educing" of old audio recordings. None of these people are responsible for any errors here.

References

Ailor, W., & Wilde, P. (2013). Re-entry operations safety. In F. Allahdadi, I. Rongier & P. Wilde (Eds.), *Safety design for space operations*. Oxford and Waltham, MA.: Butterworth-Heinemann.

Capelotti, P. J. (2010). *The human archaeology of space: Lunar, planetary and interstellar relics of exploration*. Jefferson, NC: McFarland.

Chang, K. (2012, February 18). For space mess, scientists seek celestial broom. *The New York Times*.

Chodas, P. (2002). Newly discovered object could be a leftover apollo rocket stage. Near-Earth Object Program Office. Accessed January 4, 2014, http://neo.jpl.nasa.gov/news/news134.html

Collins, L. (2012, August 6). The question artist: Tino Sehgal's provocative encounters. *The New Yorker*.

Darrin, A., & O'Leary, B. L. (2009). *Handbook of space engineering, archaeology and heritage*. Boca Raton: CRC Taylor and Francis Press.

Drew, K. (2012, March 24). Space station crew scrambles as debris passes nearby. *The New York Times*.

Feaster, P. (2012a). *Pictures of sound: One thousand years of educed audio: 980–1980*. Atlanta: Dust-to-Digital.

Feaster, P. (2012b). Extracting audio from pictures. *Media Preservation* blog, Indiana University Media Preservation Initiative. Created 20 June 2012. Accessed November 21, 2013, http://mediapreservation.wordpress.com/2012/06/20/extracting-audio-from-pictures/

Fincannon, J. (2012). Six flags on the moon. Accessed January 4, 2014, http://www.hq.nasa.gov/alsj/ApolloFlags-Condition.html

Foakes, R.A. (2005–2013). The 'Part' of Orlando in Robert Greene's Play *Orlando Furioso*. *Henslowe-Alleyn Digitisation Project*, King's College London. Accessed November 27, 2013, http://www.henslowe-alleyn.org.uk/essays/orlando.html

Fukuyama, T. (1968). *Nihon Kenchiku no Kenkyū (Studies in Japanese architectural history)*. Tokyo: Kokusui Shobō.

Gass, V., & Grosse, J. (2012, February 15). *Cleaning up earth's orbit: A Swiss satellite tackles space debris*. Accessed November 28, 2013, http://actu.epfl.ch/news/cleaning-up-earth-s-orbit-a-swiss-satellite-tack-2/

Ghassaei, A. 2012. 3D printed record. *Instructables*. Autodesk Corporation. Created 20 December 2012. Accessed November 21, 2013, http://www.instructables.com/id/3d-printed-record/

Greg, W. W. (1922). *Two Elizabethan stage abridgements: the battle of Alcazar and Orlando Furioso*. Oxford: Oxford University Press.

Hyde, J. L., Christiansen, E. L., Bernhard, R. P., Kerr, J. H., & Lear, D. M. (2001). A history of meteoroid and orbital debris impacts on the space shuttle. In H. Sawaya-Lacoste (Ed.), *Proceedings of the Third European Conference on Space Debris, 19–21 March 2001, Darmstadt, Germany*. Noordwijk, Netherlands: ESA Publications Division.

Inter-Agency Space Debris Coordination Committee Steering Committee and Working Group 4. (2007). *IADC Space Debris Mitigation Guidelines (IADC-02-01, Revision 1), September 2007*. Accessed November 21, 2013, http://www.iadc-online.org/Documents/IADC-2002-01,%20IADC%20Space%20Debris%20Guidelines,%20Revision%201.pdf

International Organization for Standardization. (2011). ISO 24113:2011: Space systems—Space debris mitigation requirements. Geneva.

Isozaki, A. (2006). Identity over time. In *Japan-ness in Architecture* (D. B. Stewart, Ed., S. Kohso, Trans.). Cambridge, MA.: MIT Press.

Jacobs, R. (2009). *Apollo: Through the eyes of the astronauts*. New York: Harry N. Abrams.

Kawamoto, S., S. Nishida, and S. Kibe. (n.d.) *Research on a space debris removal system*. Accessed November 28, 2013, http://airex.tksc.jaxa.jp/dr/prc/japan/contents/NALRP2003032/nalrp2003032.pdf

McCoy, R. (2012, January 30). No preservatives: The science and ethics of contemporary art conservation: A discussion with Tom Learner. *Art21 Magazine*.

Overhage, C. F. J., & Radford, W. H. (1964). The Lincoln laboratory west ford project—an historical perspective. *Proceedings of the IEEE, 52*(5), 452–454.

Schiffer, M. (1972). Archaeological context and systemic context. *American Antiquity*, 37.

Schiffer, M. (1987). *Formation processes of the archaeological record*. Albuquerque: University of New Mexico Press.

Slade, G. (2006). *Made to break: technology and obsolescence in America*. Cambridge, MA.: Harvard University Press.

United Nations Committee on the Peaceful Uses of Outer Space. (2007). Report of the committee on the peaceful uses of outer space. *United Nations General Assembly Official Records*, 62nd session, Supplement No. 20 (A/62/20). New York.

Vandenburg, AFB. (2012). Joint functional component command for space: Fact sheet. Accessed January 4, 2104, (http://www.vandenberg.af.mil/library/factsheets/factsheet.asp?id=12579)

Walsh, J. (2012). Protection of humanity's cultural and historic heritage in space. *Space Policy, 28*(4), 234–243.

Weeden, B. (n.d.) Billiards in space. *The Space Review*. Accessed January 4, 2014, http://www.thespacereview.com/article/1314/2

Wiedemann, C., Bendisch, J., Krag, H., Wegener, P., & Rex, D. (2001). Modeling of copper needle clusters from the West Ford Dipole experiments. In H. Sawaya-Lacoste (Ed.), *Proceedings of the Third European Conference on Space Debris* (Vol. 1, pp. 315–320). Noordwijk, Netherlands: ESA Publications Division (Published October 2001).

Chapter 7
The Preservation of California's Military Cold War and Space Exploration Era Cultural Resources

Milford Wayne Donaldson

Abstract This chapter is devoted to the research on Cold War Era properties in California, including structures devoted to research on space. Written from the perspective of an Architect and Historic Preservationist, the author focusses on the relevance to history and the challenges to preserve them. He stresses the difficulties in evaluating, interpreting and adaptively reusing historically significant buildings, structures, objects and sites. Included in the discussion are the four remaining orbiters, the *Discovery, Atlantis* and *Endeavour*, along with the proto-type *Enterprise* which, despite being less than 50 years old, are considered icons of the U.S. space program and, as such, eligible for the National Register of Historic Places.

Introduction

Department of Defense (DoD) installations in California were responsible for the research and development of Cold War communication systems, offensive and defensive weapons, and other technology which the several branches of the military used to carry out their missions and enabled the exploration of space. Military facilities of the Cold War and Space Exploration Era express the honest power and exotic nature of their operations.

But it can be difficult to evaluate, interpret and repurpose or adaptively reuse the historically significant buildings, structures, objects and sites associated with the mostly secret, but mission-critical, research, development, testing, strategic and tactical activities of the military during the period between the end of World War II in 1946 to the fall of the Berlin Wall in 1989.

During the Cold War era, the architecture and site planning of California military bases was disjointed, frequently lacked a defined style or continuity of design,

M.W. Donaldson (✉)
Architect Milford Wayne Donaldson, FAIA, Inc., Fair Oaks, CA, USA
e-mail: mwdonaldson13@yahoo.com

© Springer International Publishing Switzerland 2015
B.L. O'Leary and P.J. Capelotti (eds.), *Archaeology and Heritage of the Human Movement into Space*, Space and Society, DOI 10.1007/978-3-319-07866-3_7

and reflected three somewhat contradictory trends (Michael et al. 2011). Many properties determined to be eligible for listing on the National Register of Historic Places have been lost over the years; many more are at risk because they are not well understood, have not been identified or evaluated using appropriate historic contexts, have been deemed surplus or obsolete, or stand in the way of planning for current and future needs.

Cold War Properties of the Department of Defense

The Department of Defense (DoD) and the National Aeronautics and Space Administration (NASA) manage the largest specialized real estate portfolio in the world. The DoD currently has approximately 507,000 buildings and structures worldwide, totaling almost 2.3 billion square feet. Nearly a third of those are 50 years or older. By 2025, over 67 % of DOD's entire inventory of buildings will be more than 50 years old. This means that, including the tens of thousands of historic buildings and structures that are either now or soon to be more than 50 years old, approximately 345,000 building will potentially be eligible for the National Register of Historic Places (NRHP).

In addition to the buildings and structures, there are also a substantial number of historically significant objects like the space orbiter or testing facilities such as rocket stands (Sullivan 2006). The DoD currently manages 73 National Historic Landmarks and has nearly 600 properties listed on the National Register encompassing over 19,000 individual historic properties, located on over 200 military installations across the United States (Sullivan 2006).

Approximately 70 % of DoD's building stock 50 years or older needs to be evaluated for historical significance. As detailed in the DoD's *Cold War Needs Assessment, A Legacy Project*, completed in 2000, the identification and evaluation of the military's historic properties since 1993 under the Cold War Context had been hampered by the poorly defined or understood elements of that context due to the Cold War era's historical proximity to the present time, along with the lack of a uniform methodological approach to identifying Cold War properties, and the imposition of the standard of "exceptional importance" at the national level of significance without adequate consideration of the significance of resources at the local, regional, or statewide levels of significance (DoD 2000). As a result, many properties have been inaccurately or improperly evaluated and 'surplused' or demolished.

The need for more developed contexts and uniform tools for identification and evaluation are especially urgent with regard to military resources located in California both because of the quantity, the variety of property types, and the association of those resources with important themes in the Cold War and Space Era contexts. Although California does not lead the United States in National Historic Landmarks (NHLs), it has the most from the Cold War era (1946–1989) and the space program. California has 26 military and space-related NHLs out of a total of 137 within the state (NPS 2013). Therefore, almost 20 % of California's NHLs are military and space related, including resources from the Spanish, Russian, and American military occupation, along with space launch and testing facilities.

Driven by aggressive competition between the Soviet Union and the United States, the initial period of the early Cold War era of space exploration began with the launch of the first man-made object to orbit the Earth, the USSR's Sputnik 1, on October 4, 1957, and ended with the first Moon landing by the American Apollo 11 craft on July 20, 1969 (Logsdon and Launius 2008). Department of Defense (DoD) installations in California led many initiatives towards the research and development of those resources leading to the exploration of space.

Military facilities of the Cold War and space exploration eras express the honest power and exotic nature of their operations, but because the architecture and site planning of DoD installations in California during this period was disjointed, lacking style definition and continuity of design, it can be difficult to understand the significance of the resources, rehabilitate them for other uses, or preserve and interpret them. Security restrictions and the isolation of military installations is such that the general public is often not aware of the significance of these buildings and structures and their relationship to the development of California's historical and cultural values.

Cold War Architecture in California

During the Cold War era, from the end of World War II in 1946 to the fall of the Berlin Wall in 1989, the architecture of California military bases reflected three somewhat contradictory trends. First, for the comfort of its troops, especially the officer corps, the military built administrative and residential buildings that were commodious and fashionably modern. Second, because the military valued nimbleness and flexibility, it relied heavily upon essentially temporary buildings for most operational purposes. Third, because the military installations in California were heavily involved in weapons development and testing, rocket testing and space communications, hundreds of permanent, odd-looking buildings and structures were built to house research and test facilities.

The military adopted Modernism for administrative and residential buildings, because it was the current architectural style emerging in California and it had the sleek image of Cold War Era technology. Sometimes, military designers called upon well-known Modernist architects, such as Skidmore, Owings, and Merrill for the buildings at the Naval Postgraduate School in Monterey, or Richard Neutra, for what is still called the Neutra School in Lemoore (Moore et al. 2010). More often, the Modernist buildings were the work of lesser-known civilian architects, such as Stanley Gogerty, who laid out the modern buildings at the weapons station at China Lake, or unnamed architects who designed the beautiful Building A33 at the Bureau of Yards and Docks, Space and Naval Warfare Systems Command (SPAWAR), San Diego (Hampton 2012).

Most Cold War era buildings intended for operational purposes were designed with considerations for specific locations, to be hardened facilities capable of nuclear survivability, and meet high security requirements. They were mundane and utilitarian in appearance, designed to blend in with their environments, and were often achievements of engineering and technology. The vast majority of these Cold War era buildings, however, were built to be inexpensive and easily modified or moved. Odds are that most DoD programs were housed in some variation of a Butler Building, custom designed to meet particular needs, but engineered and constructed in a factory.

Many Cold War Era buildings and structures constructed for use as scientific or high technology facilities have undergone numerous modifications as their missions have changed and as their associated technologies have advanced. In spite of significant changes to these buildings over time, which have resulted in the loss of original design or architectural style, these buildings may be significant, not for their architecture, but for their association with the technological advancements associated with the Cold War Era.

The third category includes a huge variety of buildings and structures associated with Cold War Era weapons development and rocket testing programs, which were especially prominent in California. These were the opposite of the Butler Buildings; they were extraordinarily well-built, dedicated to specific purposes and not easily adapted for other uses. This includes the famous rocket test tracks at Edwards and China Lake; the massive rocket test stands and silos at Vandenberg and Santa Susana Field Laboratory; the great radar facilities like the Perimeter Acquisition Vehicle Entry Phased Array Warning Systems (PAVE PAWS) at Beale; and the Strategic Air Command (SAC) bomber hangars at Travis Air Force Base. In terms of Cold War era design that was historically significant and of intrinsic interest, these development and testing facilities are especially important to the historical and economic development of California.

Selected Lost Cold War Cultural Resources in California

DoD's stated policy is to manage and maintain all of its properties, not just the historic buildings and structures, through a comprehensive program that considers the preservation of their historic, archaeological, architectural and cultural values. But the need to manage these resources as mission-supporting assets and to meet the current needs of troop readiness and technological advances place many of these potentially historic facilities at grave risk for disposal, removal, and demolition as surplus properties. Preservation, interpretation, and adaptive reuse of historical properties have been complicated by two other factors. The DoD has generally made determinations of a building's useable life at around 50–70 years, and the DoD has normally received funding for new structures rather than adapting existing structures for new uses.

The following are just a handful of properties determined to be eligible for listing on the NRHP of California military facilities that have been lost over the years.

North Island Naval Air Station, San Diego, California

One of the earliest naval air stations and the first in California, North Island, San Diego became known as the "Birthplace of Naval Aviation". In 1910, a small hangar on North Island housed the shop where the concept of a drag chute for small aircraft to land safety on makeshift carriers was developed. These drag chute experiments proved to not be as safe as using sandbags with cables for arresting aircraft. The following year, in 1911, Eugene Ely made the historic flight in a Curtiss biplane on and off the USS *Pennsylvania* moored in San Francisco Bay (Pescador and Aldrich 2007). A similar cable arresting system, albeit highly advanced, is used today to safely land aircraft aboard carriers. In 1917, Congress authorized the use of North Island for training the Army Signal Corps Aviation School and the Navy Air Corps. In 1919, Hanger 17 became the first and largest structure on North Island to house a 250 feet long lighter-than-air C-6 airship. To commemorate the opening a gala party was attended by noted Hollywood stars such as Mary Pickford (Pescador and Aldrich 2007).

By 1932, the two branches were divided creating Rockwell Field for the Army and the Naval Air Station (NAS) (USDC 1934). Architect Bertram Goodhue designed NAS in 1922 (Interwar Era) in the Spanish Colonial Revival style, a style he promoted in his earlier 1915 Panama-California Exposition in San Diego. North Island Naval Air Station (NAS) provided an ideal site for aerial experiments because it had a mild climate, constant wind patterns, large areas of undisturbed land, and was away from people; a classic formula for military experiments of all kinds. Therefore, there are absolute contextual reasons why this site was chosen and the NRHP integrity issues of location, setting and association for these events become more important than the undefined architecture not associated with Goodhue's Spanish Colonial Revival style. The USN determined that the Naval Air Station comprised a historic district significant for the architecture of the buildings designed by Goodhue, but failed to address the significance of the early utility buildings or identify them as contributors to the district with the result that they have now been removed to make way for new housing, a commissary, and other troop support buildings.

USN Morris Dam Test Facility VAL, California

The Variable Angle Launcher (VAL) Test Facility was constructed in 1943 to test speed and angle of water entry of air-to-water torpedoes (Fig. 7.1). Research began after the operational failure of many aircraft-dropped torpedoes during the Battle of Midway in WWII. The Morris Dam Test Facility on the San Gabriel River, Los Angeles County, served the Department of Defense for 50 years as a test site for the recovery of underwater ordnance as well as the testing of torpedoes. This full-size launch facility used compressed air for projection and was a steel-trussed structure 332 feet long with a floating platform that allowed the angle and speed of entry of torpedoes into the water to be adjusted as needed. The test facility, along with 24

Fig. 7.1 U.S. Navy Morris Dam Test Facility, VAL, California. The Variable Angle Launcher was constructed in 1943 to test speed and angle of water entry of air-to-water torpedoes. Research began after the operational failure of many aircraft-dropped torpedoes during the Battle of Midway in WWII (U.S. Navy)

support buildings, was demolished in 1993 in a predevelopment agreement with the County of Los Angeles to return the land to its original configuration.

Building 55, Naval Air Weapons Station, Point Mugu, California

Simply known as Building 55, the modest concrete structure was built in the early 1950s as a permanent missile launching pad that replaced several scattered temporary launch sites. Although conventional, non-nuclear weapons such as missile systems were tested in California under great secrecy, the structures and buildings become obscured in their importance as a significant place where these events happened (DoD 2000).

SNORT Rocket Sled, NAWS China Lake, California

China Lake is the United States Navy's largest single landholding, representing 85 % of the Navy's land used for weapons and armaments research, development, acquisition, testing and evaluation (RDAT&E) and comprising 38 % of the Navy's land holdings worldwide. In total, its two ranges and main site cover more than

Fig. 7.2 SNORT Rocket Sled, NAWS China Lake, California. The four-mile supersonic track has created a cultural landscape, as sleds move at speeds up to Mach 4. Test tracks were also built at Edwards AFB but have been removed (U.S. Navy)

1,100,000 acres, an area larger than the state of Rhode Island. The Supersonic Naval Ordinance Research Track (SNORT) at Naval Air Weapons Station (NAWS) China Lake, formerly Naval Ordinance Test Station (NOTS), was established in 1942 (Fig. 7.2). The four-mile supersonic track created a cultural landscape, as sleds moved at speeds up to Mach 4. The hardware components are now gone; nothing remains on the dry lake bed, but the scarification on the desert floor can still be seen from the air. Test tracks were also built at Edwards AFB but have been removed entirely (DoD 2000). An image of the water braking system in action can only be seen on film. NAWS is also the home of Cosco Rock Art, a 12,000-year-old archaeology site and a National Historic Landmark.

Arctic Submarine Laboratory, Battery Whistler, Point Loma, California

The 1916 Battery Whistler, a mortar emplacement, was officially decommissioned in 1943; in 1947 the Navy began using the site as a submarine research facility. The Arctic Submarine Laboratory (ASL) was a research facility of the U.S. Navy's Electronics Laboratory (NEL) (Fig. 7.3). Created by Dr. Waldo Lyon, the facility, housed in a corrugated metal utility building, contained a pool equipped to freeze salt water and to grow sea ice to study their physical properties on submarines. In 1960, Dr. Lyon was awarded Distinguished Federal Civilian Service Medal by President Kennedy.

Fig. 7.3 Arctic Submarine Laboratory, Battery Whistler, Point Loma, California. Created by Dr. Waldo Lyon, the facility, housed in a corrugated metal utility building, contained a pool equipped to freeze salt water and to grow sea ice to study their physical properties on submarines. Research culminated in the USS *Nautilus,* the first nuclear powered submarine to travel 1000 miles, taking 74 h underneath the North Pole in 1958 (U.S. Navy)

Research at the ASL culminated in 1958 with the USS *Nautilus*, the first nuclear- powered submarine, completing a transpolar submerged voyage of 1,000 miles in 74 h. The Arctic Submarine Lab also included a sea ice cryostat for testing scale-model submarine sails designed to punch up through the ice, and possessed the only high-energy electron-producing betatron on the West Coast, used for examining the structures of heavy objects and metals up to 18 inches in diameter.

Through the 1970s and 1980s, the ongoing research at the ASL resulted in refurbishment and improvement of the Lab's cryogenic facilities. These facilities were used for evaluating icing issues on *Los Angeles* class submarines, sonar technology developments for remote acoustic measurement of ice thickness, and the ice breakthrough tests for *Seawolf* class submarines. The Arctic Submarine Laboratory was demolished in 1996, two years before the death of Dr. Lyon, to expose the remains of the 1916 Battery Whistler, a resource deemed to be more important by the USN.

NTS/NSY Roosevelt District, Long Beach, California

Knowing that United States entry into WWII was imminent, President Franklin D. Roosevelt set out to increase naval training facilities on the West Coast. Work began in 1940 at the Naval Training Station and Naval Ship Yards (NTS/NSY) with a schedule for opening in 1945 as home of the Pacific Fleet (Fig. 7.4).

Fig. 7.4 NTS/NSY Roosevelt District, Long Beach, California. President Franklin Delano Roosevelt, knowing that US involvement into WWII was imminent, set out to increase training facilities on the West Coast. Work began in 1940 with a schedule for opening in 1945 as home of the Pacific Fleet. Home to 25,000 civilian and military personnel and serving almost 700 battle-scarred ships during WWII (photo by the author)

Designed by Paul Williams, a well-known African-American Los Angeles-based architect, in the International Style and built of reinforced concrete by Allied Engineers and Architects, the master-planned campus was home to 25,000 civilian and military personnel and served almost 700 battle-scarred ships during WWII.

As a result of the 2005 Defense Base Closure and Realignment Commission (BRAC) action the entire facility, including never occupied newly constructed enlisted men barracks, was demolished to make way for the expanded Port of Long Beach and the San Pedro Container Port.

California Naval Facility Centerville Beach

Naval Facility Centerville Beach is located in rural Humboldt County along the Pacific Ocean coast, about five miles west of the small village of Ferndale. The 32-acre facility opened in 1958 as part of the Navy's secret world-wide Sound Surveillance System (SOSUS). The program was established to conduct underwater acoustic surveillance of notoriously loud Soviet submarines during the Cold War Era. Its successes were numerous, including detection of a submarine in the Caribbean Sea during the Cuban Missile Crisis.

Architecturally, the Humboldt County facility was modest, exhibiting typical Cold War era military construction. As the Cold War era ended and improved submarine technology led to quieter vessels, SOSUS became nearly obsolete. Subsequently, in 1993, most of the program facilities closed and were eventually demolished, including Naval Facility Centerville Beach. A historic resources report was prepared for Naval Facility Centerville Beach to assist the Navy's compliance with Section 106 of the National Historic Preservation Act (Herbert and Freeman 2009).

Limited Successes on Rehabilitating Cultural Resources

The following are examples of encouraging efforts by the DoD to preserve portions of significant cultural resources associated with the Cold War and space exploration eras. Several of these projects are still in process but the final outcome is promising.

Chollas Heights Naval Radio Transmitting Facility, San Diego, California

Constructed during the Modernization Era in 1917, the Chollas facility was the largest and most powerful radio transmitter in North America (Fig. 7.5). It included three

Fig. 7.5 Chollas Heights Naval Radio Transmitting Facility, San Diego, California. Constructed during the Modernization Era in 1917, the Chollas facility was the largest and most powerful radio transmitter in North America. It included three 600-foot tall towers with a copper antenna suspended mid-way between the towers. The facility was the first in the development of long-range transmitter stations between Arlington, Virginia; Pearl Harbor, Hawaii and Cavite, Philippine Islands (U.S. Navy)

600-foot tall towers with a copper antenna suspended mid-way between the towers. The facility was the first in the development of long-range transmitter stations between Arlington, Virginia; Pearl Harbor, Hawaii; and Cavite, Philippine Islands. This high-tech communication system replaced naval ships strategically spaced at sea so as to be able to transmit to one another due to the curvature of the earth. The Navy departed from their East Coast Colonial Heritage styles and introduced their new Mission Revival designed buildings, along with Craftsman and Island architectural styles.

As a credit to the Navy, the buildings were rehabilitated into a community center and many artifacts were saved and interpreted within the facility. Unfortunately, the towers and antenna, district contributors and the most historically significant features of the facility, were demolished.

Hangar I, Moffett Field, Sunnyvale, California

Hanger 1 in Sunnyvale, near San Jose, was built to house the airship USS *Macon*. The 17-story tall, 361,000 square foot Streamline Moderne style hangar is large enough to hold seven football fields. Designed by Karl Arnstein of the Goodyear Zeppelin Company and built in 1933, Hangar 1 is a contributing element to the United States Naval Air Station Sunnyvale Historic District. Arnstein also designed the Akron, Ohio Hangar (1930), the USS *Akron* (1931), and the USS *Macon* (1933), both LTA helium-filled airships. After it was discovered that the building was leaching polychlorinated biphenyls (PCBs) into the Moffett storm-water settling basin groundwater, the Navy proposed a Comprehensive Environmental Response, Compensation, and Liability Act (CERCLA) Non-Time Critical Removal Action.

The Navy considered 13 alternatives for hazard mitigation and was favoring demolition. At the urging of several California Congressional members, preservationists statewide, the California State Historic Preservation Officer (CA SHPO), the National Trust for Historic Preservation, and the Advisory Council on Historic Preservation (ACHP) as consulting parties under Section 106 of the National Historic Preservation Act (NHPA), an alternative to demolition was agreed upon by the Navy in 2010, whereby the hangar would be transferred to NASA (NHPA 1966). The action included removal of the original siding, deconstruction of original interior structures, removal of debris to appropriate off-site disposal or recycling, cleaning by high-pressure washing and/or other mechanical means, and application of an epoxy coating system to the Hangar's remaining structural steel frame.

After transfer, NASA will be responsible for re-siding the hangar matching the original siding, allowing for the survival and reuse of this engineering marvel. As a young engineer testified during the ACHP public hearing in Sunnyvale and reading passionately from his glowing laptop at the podium, "Hangar 1 is our icon, our landmark in Silicon Valley, our dot-com from the 1930s and a testament to American ingenuity." Another local supporter asked, "Would the Statue of Liberty, stripped of its copper cladding, standing naked as a frame, continue to be a symbol of American freedom?" (Donaldson 2008).

Pioneer Deep Space Station, Mojave Desert, California

The Goldstone Deep Space Communications Complex (GDSCC), commonly called the Goldstone Observatory, is located in California's Mojave Desert. Operated by International Telephone and Telegraph (ITT) Corporation for the Jet Propulsion Laboratory, its main purpose is to track and communicate with space missions. It includes the Pioneer Deep Space Station, which is a National Historic Landmark. Constructed in 1958, the Pioneer Deep Space Station was the first antenna to support the National Aeronautics and Space Administration's unmanned exploration of deep space, and the prototype antenna for the entire Deep Space Network for tracking deep space vehicles. Goldstone antennas have also been used as sensitive radio telescopes for such scientific investigations as mapping quasars and other celestial radio sources; radar mapping planets, the Moon, comets and asteroids; spotting comets and asteroids with the potential to strike Earth; and the search for ultra-high energy neutrino interactions in the moon by using large-aperture radio antennas.

The Pioneer Deep Space Station is scheduled to be dismantled, moved to a new site and reassembled by the Barstow Community College Space and Technology Center, a Smithsonian Institute Regional Museum, as a visual centerpiece artifact on campus. Listed on the NRHP in 2007, it will lose its status as a NHL, since its original location, settling, and site association with the GDSCC will no longer exist (NPS 1999).

Edwards Air Force Base Historic Context Statement on Cold War

In an effort to cope with an ever-increasing scarcity of resources and funds, Edwards Air Force Base has developed a proactive, broad-based management strategy to help focus limited assets on more productive historic preservation initiatives, and has developed a *Historic Context Statement Report for Evaluation of Cold War Era Properties on Edwards Air Force Base, California.*

The study presents a comprehensive and well-defined plan to adequately identify historic district boundaries on Edwards AFB, as well as systematic methods to differentiate between properties that are eligible or ineligible for listing on the National Register of Historic Places. Rather than assessing buildings in contextual isolation, they aim instead to implement management strategies that focus on evaluating and protecting potential historic districts. The development of a historic context is not considered an undertaking but will be used in identifying potential cultural resources and evaluating the effects of proposed undertakings on the base's historical resources.

Vandenberg Air Force Base

Vandenberg Air Force Base (VAFB), the third largest Air Force installation, is home of the 30th Space Wing. Remotely located on California's Central Coast, the installation provides America's only capability to launch military and commercial satellites into polar orbit and conduct intercontinental ballistic missile testing without over-flying populated areas. The 30th Space Wing also operates the Western Range consisting of instrumentation sites along the California coast providing a vast array of space and missile tracking equipment and is also home to Missile Defense Agency test and operations programs.

The Solid Controlled Orbital Utility Test (SCOUT) complex, in operation between 1961 and 1994, was used primarily to launch the SCOUT series of NASA-designed solid-fuel rockets that placed small research satellites into orbit. The central piece of launch hardware at Space Launch Complex 5 (SLC-5), the SCOUT erector/launcher, is one of only three in the world.

Unfortunately, SLC-5's erector portion was demolished in 2011 due to obsolescence. The SCOUT launcher was salvaged for re-development at White Sands and re-deployment to the Pacific Missile Range Facility where the old SCOUT hardware will find new life in a program exploring rail-launched space vehicles capable of launching small satellites for technology demonstration missions at low cost.

The Memorandum of Agreement (MOA) between Vandenberg AFB and the California SHPO called for the production of a Historic American Engineering Record of the complex as mitigation for the adverse effects of its dismantling. Transfer and adaptive reuse of this rare Cold War Era erector/launcher made a new launch mission possible, resulted in $500 K savings over the cost to build new launch hardware and $2 M in overall program cost avoidance (DoD 2008).

Vandenberg also successfully completed the final phase of an 8-year, $4 M project for preservation of Space Launch Complex 10 (SLC-10), a National Historic Landmark built in 1958 for the U.S. Air Force's Intermediate Range Ballistic Missile training and testing program, and later adapted for space flight purposes (NPS 2013). These preservation and rehabilitation efforts ensure that SLC-10 will remain the best surviving example of an early launch complex built during the infancy of U.S. space exploration and reconnaissance programs. SLC-10, the Missile Heritage Center, preserves one of the most extensive collections of launch control hardware from the 1950s.

Shuttle Components, Crawler, Launch Pad Facilities, Kennedy Space Center, Florida

The Space Transportation System (STS), better known as the Space Shuttle, was established in the late 1960s to create reusable space vehicles that would enter space, return to earth, and prepare for another flight. The program commenced on April 12, 1981, with *Columbia* and STS-1, the first shuttle orbiter flight.

A total of six shuttle orbiters were built, the *Enterprise*, a prototype used for gliding tests only and the other five for spaceflight soon became icons of the U.S. space program. The first two, *Columbia* and *Challenger*, met tragic ends, but the remaining three, *Discovery, Atlantis,* and *Endeavour*, continued their epic service until retirement in 2011. The final mission of the space shuttle program was STS-135 flown by *Atlantis*, in July 2011. Despite being less than 50 years old, NASA determined the four shuttles were eligible for the National Register of Historic Places for their outstanding contribution to space flight and exploration and significant engineering. With the exception of the *Enterprise*, each orbiter's three main engines, external tank, and solid rocket boosters are contributing historic elements.

Through the NHPA Section 106 process, NASA, the ACHP, and the SHPOs from Texas, Florida, Alabama, and California worked to ensure the exciting story of this program and the contributions made to space travel would be preserved and told in various formats reaching broad audiences such as school children, the public, scientists, and space professionals. In addition, support equipment and facilities are being recorded, providing permanent textual and visual documentation of the entire STS. The remaining four shuttles are now preserved, interpreted, and on display; *Discovery* is at the Smithsonian Institution's National Air and Space Museum Udvar-Hazy Center in Virginia; *Endeavour* is at the California Science Center in Los Angeles; *Atlantis* is now at the Kennedy Space Center in Florida, and the *Enterprise* is at the Intrepid Sea, Air & Space Museum in New York City. Testing, assembly, maintenance, and launch facilities are being evaluated for potential new uses, but many are obsolete for NASA's ongoing missions and are slated for demolition (ACHP 2012).

Potential Loss of Cultural Resources and Projects to Watch

The following projects are in critical phases of development and many cultural resources have been determined to be eligible for listing in the NRHP. The outcome of these projects at this time is unknown. These projects also show that the general public is not as involved as with some of the earlier, more successful preservation efforts, and serve as testimony to how vulnerable these cultural resources are when located on isolated and secure sites.

Naval Weapon Station Seal Beach, Demolition of NASA Saturn S-II Historic District

The Navy proposes to demolish buildings 112, 126 and 127, and consolidate the operations in a newly constructed one-story Strategic Systems Weapons Evaluation Test Lab. The existing buildings are noted by the Navy to be deteriorating and unsuitable. The buildings, used in the 1960s and early 1970s for the final assembly and test of second stage boosters for the Saturn V moon rocket, have been determined to be eligible for listing on the NRHP as contributors to NASA Saturn S-II Historic District, identified in 1999 in an inventory and evaluation of 178 Cold War Era properties at Seal Beach. Twenty-two properties appear to meet the criteria for eligibility as contributors to the district. The Navy's plan for the area includes eventual demolition of the remainder of all the resources in the eligible district, thereby rendering the NASA Saturn S-II Historic District ineligible for NRHP listing (Herbert and Freeman 2009).

PAVE PAWS Installation, Beale AFB, California

Built in 1977, the Perimeter Acquisition Vehicle Entry Phased Array Warning Systems (PAVE PAWS) at Beale AFB, near Sacramento, is a very rare 10-story facility directly associated with the late Cold War era. This facility, a component of the Air Force Space Command Radar System, was linked with facilities located at Clear Air Force Station in Alaska (AFS) and on Cape Cod in Massachusetts to detect and track sea-launched missiles and Intercontinental Ballistic Missiles (ICBMs). These three sites communicated with each other and relayed the information to the Cheyenne Mountain Air Station using phased array antenna technology.

The phased array antenna radar systems differ from mechanical radars, which must be physically aimed at an object for tracking and observation. The phased array antenna remains in a fixed position. Phased array antenna aiming, or beam steering, is done in millionths of a second by electronically controlling the timing, or phase, of the incoming and outgoing signals. The Beale AFB PAVE PAWS Radar was recently upgraded to an Upgraded Early Warning Radar to support the co-primary Missile Defense Mission. The Cape Cod radar is currently being upgraded (Donaldson 2012).

The fate of these types of facilities is unknown as technology changes so rapidly. This can also be said of NASA's enormous wind tunnels at Moffitt Field. Broader worldwide Cold War Era contextual studies are needed to understand these cultural resources (NPS 1999).

Santa Susana Field Laboratory

After WWII, Rockwell established the Santa Susana Field Laboratory in the Simi Hills, Simi Valley of Southern California to test engines for missiles, spacecraft, and rockets in the Santa Susana Mountains, where actor Tom Mix once filmed silent westerns. Selection and design of the site was influenced by the German WWII test program, which had done most of its V-2 rocket testing at abandoned rock quarries in Germany. Santa Susana's natural bowl area and canyons were very similar and the expatriate German scientists assisting the Americans knew how to use them.

After construction began in 1947, the Santa Susana Field Laboratory location was used by a number of companies and agencies (Fig. 7.6). The first was Rocketdyne, originally a division of North American Aviation (NAA), which developed a variety of pioneering, successful and reliable liquid-propellant rocket engines such as those used in the Navaho cruise missiles, the Redstone rockets,

Fig. 7.6 Vertical Test Stand One, Santa Susana Field Laboratory, California. The Santa Susana Field Laboratory is a complex of industrial research and development facilities located on a 2,668 acre portion of the Southern California Simi Hills in Simi Valley, California, used mainly for the testing and development of liquid-propellant rocket engines for the United States space program from 1949 to 2006 (photo by the author)

the Thor and Jupiter ballistic missiles, early versions of the Delta and Atlas rockets, the Saturn rocket family, and the Space Shuttle main engines. The Atomics International division of NAA utilized a separate portion of the Santa Susana Field Laboratory to build and operate the first commercial nuclear power plant in the United States and for the testing and development of compact nuclear reactors, including the first and only known nuclear reactor launched into Low Earth Orbit by the United States, the SNAP-10A.

The Santa Susana Field Laboratory includes sites identified as historic by the American Institute of Aeronautics and Astronautics and by the American Nuclear Society. In 1996, the Boeing Company became the primary owner and operator of the Santa Susana Field Laboratory and later closed the site (NASA 2010a).

The 2007 Historic Resource Assessment Survey included an initial review of a list of 135 NASA owned buildings, structures, and sites located within the Santa Susanna Field Laboratory. After archival study and field research, six test stands located in the Alfa, Bravo and Coca test areas, plus three associated control houses were evaluated as meeting National Register criteria of eligibility within the contexts of the Cold War Era and Space Exploration, circa mid-1950s to 1991. In addition to the nine individually eligible historic properties, three historic districts were identified as eligible for listing in the National Register: The Alfa Test Area Historic District, the Bravo Test Area Historic District, and the Coca Test Area Historic District. Each is considered eligible within the contexts of the Cold

War Era and Space Exploration. The relevant areas of significance are Military, Engineering, Transportation, and Space Exploration (NASA 2010b).

The Bravo Test Area Historic District (BTA) contains eight contributing resources and one noncontributing resource. Constructed during 1955–1956, the Bravo test site featured the second cluster of static test stands operational for AFP 57 at Santa Susana. The engineers relied on primitive test gear, using vibration monitors and oscilloscopes cannibalized from oil-drilling companies. What gauges they lacked to measure horrifically strong flame and thrust, they built from scratch.

A turning point came in 1950 with the first successful test of the Redstone, a V-2 offspring carrying America's first nuclear warheads and, in 1961, when Mercury astronaut Alan Shepard blasted off in the first manned US rocket flight. The Bravo Test Area is considered eligible under Criterion A for its associations with multiple static engine tests run between 1956 and 1991, beginning with tests of Atlas thrust chambers in 1956, and also supporting testing of F-1 components, Lunar Module Rocket Engine assemblies, as well as Atlas and Delta RS-27 Vernier engines and turbo pumps. The Bravo Test Area Historic District is also significant under Criterion C for the design and engineering of the test site by Daniel, Mann, Johnson and Mendenhall (DMJM), with Walter Riedel; district contributors include the test stands and blockhouse, the ancillary buildings and structures, and elements of the natural and man-made landscape (NASA 2010b).

Under an Administrative Order on Consent with the State of California's Department of Toxic Substances Control (DTSC), NASA has been tasked with cleanup of toxic materials at the SSFL to a "background level" that far exceeds other, similarly mandated cleanups. This means that NASA will have to remove in some places as much as 30 feet deep of contaminated soil, and replace it with dirt brought in from outside the area. NASA is in discussions with the State of California about the exact levels the facility will need to be cleaned to.

Cleanup to this high standard could mean that several archaeological sites would be affected if the underlying soil were found to be contaminated. NASA proposes to demolish the most contaminated test stands in the Coca historic district in its entirety as this has the largest test stands and the most extensive contamination. It is also closest to the core of the Indian Sacred Site declared by the Santa Ynez Band of Chumash Indians, a consulting party; the Santa Ynez Band considers this entire area to be a Traditional Cultural Property (NASA 2010b).

In a recent discussion with ACHP staff in August 2013, at a minimum, NASA intends to retain one test stand and control house, and encapsulate any contamination and protect these structures with a fence. NASA will request that GSA put a covenant on the deed to protect the test stand and control house. NASA will preserve artifacts for display from test stands that were demolished.

Total cleanup of Santa Susana is slowly moving towards returning the site to original open space. There are those, however, who harbor the wish that some remnants of the site's exciting history will be left for future generations to visit and ponder. The site played, after all, a central role in this nation's race to develop vehicles to assure America's dominance in space. Santa Susana and sites like it played some of the most important roles in Cold War Era and manned space flight history.

Conclusions

Department of Defense services began identifying and evaluating military facilities associated with the Cold War Era and Space Exploration in the early 1990s. In accordance with DoD and NASA guidance, the focus of historic context development has been at the national level. While that approach has considerable logic, its results have met with mixed success; reviewing agencies responsible for local or state level contexts and interested parties and associations with local, regional, or industry interests have been critical and/or rejected a number of federal agency reports.

Historical contexts focused only on the national significance of Cold War military properties miss the significance of those to the growth and developments of local communities and regions such as those in California where the lives of individuals living and working within communities that provided services to the bases and military installations were shaped by the military programs and installations both during and after the Cold War. It would appear that consideration of state and local impacts is primarily encouraged by agencies in states with few Cold War era or space exploration resources, while state offices reviewing numerous of these facilities appear to consider only national significance in the context of the history of the Cold War era or space exploration periods (Meltzer 2011).

The rationale for the preparation of the historic context requires an understanding of the significance of both the broad and specific historical events and persons associated with a property or installation and how those events and persons are associated with the properties being evaluated (King 2011). This understanding must be appreciated and understood by both the investigator and the reviewer.

When the application of the Cold War era and space exploration focused or site- specific historic contexts by federal agencies became prevalent, a number of broad, national-level historic contexts were also prepared. Among others, these include the general Cold War era and space exploration related contexts prepared by NASA, the U.S. Navy, and Air Force, as well as specific contexts relating to a variety of subjects such as guided missiles, communications and radar systems, and defense production during the Cold War era and space exploration. Both government and privately contracted historians have prepared these contexts.

To meet the criteria of 'exceptional importance' required for properties less than 50 years in age, temporal associations with these broad contexts is not enough. As a result, additional, more focused historical research is always required for each property or installation being evaluated. The additional research ensures that properties are considered within the broadest possible range of contexts, an element of the process that is particularly important in cases involving highly specialized missions, like space exploration, or where state or local associations are present and public sentiment and emotion may be involved. In the case of the latter, some military installations, in particular remote facilities, are often inextricably tied to a community and its economy. However, this is not necessarily a historically significant relationship (King 2011).

Recently, the Department of Defense made a presentation to the Advisory Council on Historic Preservation council members on their study on sustainability and DoD's historic pre-World War II masonry buildings. The study concluded that these buildings are valuable assets that can be modernized to match the energy performance of new buildings at significantly less cost than new construction. However, such benefits may be negated if antiterrorism/force protection and progressive collapse standards are applied in a rigid and prescriptive manner.

If the tangible history of these rare and important buildings, structures, objects and sites is to be preserved, we must develop a better understanding of their designs, engineering, and of the roles they played in the history and development of our communities, states and nation. We need to create a centralized repository and data clearinghouse. We need to develop better, more useful, historic contexts that look beyond simply architectural significance of individual properties, uniform evaluation methodologies, and a more refined understanding of the significance of a broader range of property types associated with important themes within those contexts. We need to develop more creative and improved mitigation and treatment options, and use updated historic building cost/benefit analyses for adaptive reuse considerations. And we need to develop best management practices for Traditional Cultural Properties, and improve tools for identifying and evaluating cultural resources in inaccessible areas (Smythe and York 2009).

As we move towards the 50th celebration of the National Historic Preservation Act of 1965, we need to recapture the sense that increased knowledge and preservation of the nation's irreplaceable historic resources is important for the economic growth and development of our communities, regions, and states, as well as the nation as a whole.

We need to reaffirm the principles on which the National Historic Preservation Act was based, that stewardship of historic resources inspires and benefits present and future generations.

As we move forward in protecting our history through our preservation of these military and space related resources, it is not sufficient to leave only memories, perceptions or expressions captured in an obscure report filed away from public recognition. These tangible assets of the Cold War and space exploration eras, albeit difficult to reuse but none the less significant, must be preserved as icons for what they can teach us and future generations about the very special period during the nation's and the world's history when we faced both the threat of nuclear holocaust and celebrated the incredible achievement of putting a human being on another celestial body. To do this, we must preserve these extraordinary cultural resources of the Cold War era and space exploration period for generations to come as our legacy.

References

Advisory Council on Historic Preservation (ACHP). (2012). *106 success story space shuttles: NASA contribution to space flight and exploration.* Accessed June 22, 2013, http://www.achp.gov/docs/Section106SuccessStory_NASAShuttles.pdf

Department of Defense. (2008). *Secretary of defense environmental awards, Vandenberg air force base*, California." Accessed January 12, 2013, http://www.denix.osd.mil/awards/upload/2_Vandenberg_Air_Force_Base-_California.pdf

Department of Defense. (2000). *Cold war needs assessment: A legacy project.* Accessed January 6, 2014, http://www.denix.osd.mil/cr/upload/LRMP_98-1754_Cold_War_Needs_Assessment_2000.pdf

Donaldson, M. W. (2012). Countdown to disaster: Perspectives in the preservation of cold war era cultural resources. *Preservation Matters 5*(2), California Office of Historic Preservation. Accessed March 13, 2013, http://ohp.parks.ca.gov/pages/1054/files/corrwinter2010.pdf

Donaldson, M. W. (2008). Save hangar one!!. *Preservation Matters 1* (4), California Office of Historic Preservation. Accessed March 13, 2013, http://ohp.parks.ca.gov/pages/1054/files/fall%20%20november1%20newsletter_2008.pdf

Hampton, R. (2012). *Historic context for evaluating mid-century modern military buildings.* Washington DC: Department of Defense: DoD Legacy Resource Management Program.

Herbert, R., & Freeman, J. (2009). *Naval facility centerville beach historic resources inventory and evaluation.* Report by JRP Historical Consultants LLC prepared for Naval Facility Southwest Division Engineering Command. Davis, California.

King, T. (Ed.). (2011). *A companion to cultural resource management.* West Sussex, UK: Wiley-Blackwell.

Logsdon, M., Launius, R. (Eds.) (2008). *Exploring the unknown, selected documents in the history of the U.S. civil space program,* Vol VII, Human Space Flight: Projects Mercury, Gemini and Apollo. Washington, DC: NASA SP: 4407.

Meltzer, M. (2011). *When biospheres collide: A history of NASA's planetary protection programs.* Washington DC: U.S. Government Printing Office.

Michael, M., Smith, A., Sin, J. (2011). *The architecture of the department of defense, a military style guide.* Washington, DC: Department of Defense, DoD Legacy Resource Management Program.

Moore, D., Edgington, J., Payne, E. (2010). *A guide to architectural and engineering firms of the cold war era.* Washington DC: Department of Defense: DoD Legacy Resource Management Program.

NASA. (2010a). *Santa susana field laboratory, an overview of NASA's history at SSFL.* Washington DC: SSFL-2010-04-27ADM.

NASA. (2010b). *A look back at space engine testing at the santa susana field laboratory.* Washington DC: SSFL-2010-04-26.

National Historic Preservation Act (NHPA). (1966). *1966 as amended through 1992. Public Law 102–575.* Washington DC: Government Printing Office.

National Park Service (NPS). (2013). *List of national historic landmarks.* Washington DC: Department of the Interior, National Park Service. Accessed September 22, 2013, Cr.nps.gov/nhl/designations/listsofNHLs.htm

National Park Service (NPS). (1999). *Guidelines for completing national register of historic places forms part b how to complete the national register multiple property documentation form.* Washington, DC: U.S. Department of the Interior.

Pescador, K., & Aldrich, M. (2007). *Images of aviation, San Diego's North Island 1911–1941.* San Francisco, California: Arcadia Publishing.

Smythe, C., & York, F. (Eds.) (2009). Traditional cultural properties: putting concept into practice. The national register framework for protecting cultural heritage places. *The George Wright Forum 26*(1) 18–19.

Sullivan, M. (2006). Current state of DoD historic properties. In *DoD Cultural Resources Workshop on Prioritizing Cultural Resource Needs in Support of a Sound Investment Strategy,* Seattle, Washington, July 11–13, 2006.

U.S. Department of Commerce (USDC). (1934). *Descriptions of airports and landing fields in the United States.* Washington DC: Bureau of Air Commerce.

Chapter 8
Legal Implications of Protecting Historic Sites in Space

Joseph Reynolds

Abstract This chapter, written by a recent graduate of Clemson University/The College of Charleston with an M.S. in Historic Preservation, examines the legalities of preserving the heritage of space exploration. The chapter briefly describes the unique social and political environment and historical context of lunar sites and artifacts in terms of preservation. It uses case histories to demonstrate the complexities involved with using legal structures to protect them in the future.

Introduction

This chapter examines whether the international legal regimes created during the space race of the 1960s will allow for the protection of the remains of one of mankind's most significant achievements. Since all of the objects that were left on the Moon during the Apollo missions are still U.S. government property, this text also explores U.S. preservation law. The ultimate goal of this endeavor is to have the Apollo lunar landing sites included onto the World Heritage List so that they will achieve a global level of commitment to preservation.

In order to fully understand the rules and regulations that pertain to the Moon, outer space, and other celestial bodies, this chapter examines the work of legal historians and space enthusiasts in regards to the major space treaties. NASA documents, National Park Service archives, various World Heritage Sites, scientific journals, and news outlets were combed to provide insight, information, and historical data on the topic. Because outer space and its celestial bodies have been deemed areas of international commons, international conventions pertaining to the ocean floor and Antarctic continent were examined for similarities.

J. Reynolds (✉)
Portland, ME, USA
e-mail: Reynolds.josephp@gmail.com

© Springer International Publishing Switzerland 2015
B.L. O'Leary and P.J. Capelotti (eds.), *Archaeology and Heritage of the Human Movement into Space*, Space and Society, DOI 10.1007/978-3-319-07866-3_8

The Preservation of Historic Sites in Space

When Astronauts Neil Armstrong, Edwin "Buzz" Aldrin, and Michael Collins returned to Earth safely from the first successful journey to the Moon on July 24, 1969, they did not return to find the laws of the United States, or the international community interested in forever preserving the site of their historic achievement. Instead, the site in which human beings first set foot on another celestial body has been preserved by an entirely different method, the vacuum of space.

The Apollo landing sites are not only significant because of their importance to human scientific achievement but also because they are the only sites in human history that have sat frozen in time. The Apollo 11 Landing Site consists of 106 objects made specifically for the first manned mission to the Moon's surface including the lunar module lander, active NASA experiments, and humanity's first footsteps on the Moon. This site is roughly the size of a baseball diamond and constitutes the first archeological site with human activity on another celestial body.

The lack of atmospheric conditions on the Moon has created an almost perfectly preserved site because it has dealt with little interference since Armstrong and Aldrin left it in 1969. The extremely delicate nature of that site creates a very difficult situation in regards to protection and preservation if and when humans should return, especially when the site is in danger from the next wave of potential lunar explorers.

The space tourism industry is not as far off as people believe it to be. Companies like Virgin Galactic are within years of being able to bring travelers into space and eventually the Moon. The Google LunarX competition is offering a $30 million prize to the first team of scientists to design a rocket that will get a robotic spacecraft to the Moon. In order to prevent the potential human disturbance of this historic site this chapter explores the idea of preserving human artifacts 239,000 miles away.

A Brief History of Space Laws and How They Interact
with Historic Preservation

The legality of preserving human archeological sites on the moon is a complex topic that deals with some universally accepted, and some highly controversial international laws. Because the objects located on the lunar surface are still considered United States Government property it is important to examine how US law can provide protection.

The United Nations Committee on the Peaceful Uses of Outer Space (COPUOS) drafted the five major space treaties governing human usage of the heavens from 1967 to 1979. COPUOS was an ad hoc committee created by the UN to explore "the nature of legal problems which may arise in the carrying out of programs to explore outer space (Hosenball 1979)." The committee was created in 1958, 1 year after the launch of *Sputnik*, and within weeks of the creation of NASA. COPUOS was formed with 24 members, only 1/3rd of the members of the committee today (UN Committee on Peaceful Uses of Outer Space 2001). The members are split into two sub-committees, one being the Legal Sub-committee, comprised of lawyers, legal scholars, and diplomats; and the Scientific and Technical Sub-committee, comprised of members with scientific backgrounds.

Members of the early sessions of COPUOS were drafting language to govern outer space and its celestial bodies without any precedent and, in some instances, prior to the first humans stepping foot on the Moon. Laws pertaining to space had never been created before, and so these legal pioneers had to draw from more theory and less hard data. As a result, omissions and loopholes were bound to plague these new treaties, not the least of which pertains to historic preservation.

The foundation of international space law is *The Treaty on Principles Governing the Activities of States in the Exploration and Use of Outer Space, Including the Moon and Other Celestial Bodies*. What is commonly referred to as the Outer Space Treaty (OST 1967) is the first of five space laws created by the Legal Sub-Committee of the United Nations Committee on the Peaceful Uses of Outer Space (UN Committee on the peaceful uses of outer space 2001). While it is not the first treaty passed in regards to human usage of the heavens, the OST is an advancement of the principles set forth in previous general assembly resolutions, international agreements, statements by elected government officials, domestic laws, and the opinions and articles written by scholars in the field.

Largely written over the summer of 1966 during the fifth session of the UN COPUOS, the creation of the majority of the treaty's text was by US and Soviet delegations that wanted an agreement before the first human landed on the Moon. The drafting of this treaty was a historic moment in international law because two nations with radically different political ideologies, put their differences aside and compromised on a treaty that would help structure the next age of human exploration.

The Political Committee of the UN General Assembly approved a first draft of the OST on December 17, 1966, and it was endorsed unanimously by the General Assembly 2 days later. The OST was opened for signature in London, New York, and Moscow in January of 1967 and entered into force in October of that year. The OST was not the first legally binding document to curtail human behavior in space. As early as 1959, the American Bar Association (ABA) passed a resolution declaring, "in the common interest of mankind…celestial bodies should not be subject to exclusive appropriation (Dembling and Arons 1967)." Resolutions like this mirror how edgy the US was about losing to the USSR in a race to the moon.

In 1965–1966 the UN was pushing COPUOS to create a treaty intensifying on the principles of *The Declaration*. Specifically, the General Assembly was urging COPUOS to draft internationally binding legislation on the issues of assistance and return of astronauts, liability, and the exploration and uses of outer space. Not only did COPUOS have the General Assembly to deal with, on May 7, 1966, US President Lyndon B. Johnson stated "the need to take action now…to insure that explorations of the moon and other celestial bodies will be for peaceful purposes only" (The Miller Center, University of Virginia 2012).

On June 16, 1966, US Ambassador to the UN Arthur J. Goldberg submitted the American draft of *The Treaty Governing the Exploration of the Moon and Other Celestial Bodies* to the Chairman of COPUOS. On that same day, Platon Morozov, the Soviet Ambassador to the UN, submitted the Soviet draft of "Treaty Principles Governing the Activities of States in the Exploration and Use of Outer Space, the Moon, and Other Celestial Bodies." COPUOS agreed to begin drafting the discussions on the treaty on July 12, 1966, the date the US wanted to begin, in Geneva, Switzerland, the place the Soviets wanted it to take place (Dembling and Arons 1967).

During the Fifth Session of UN COPUOS, the 28 member delegation decided that a decision needed to be made rapidly on the rules of conduct of states on celestial bodies due to the impending landing of humans on the moon by either the US or Soviet governments. There was an overwhelming agreement by the delegates that the use of celestial bodies for military purposes, especially weapons of mass destruction, was not desirable for the future of our species and should be forbidden with this treaty.

The OST is composed of 17 articles that accompany UN General Assembly Resolution 2222. An article-by-article analysis of the OST and how it relates to the preservation of the Apollo Lunar Landing Sites would be necessary for a complete understanding of the topic. For this chapter only the articles of the treaty that are necessary will be discussed.

Articles I, II, and III of the OST address the issue of land claims on celestial bodies by sovereign states; the OST strictly forbids any state from claiming or appropriating land on the surface of any celestial bodies for the purposes of expanding a state's territory, economic claims, or discovery of mineral resources. It states that the exploration of outer space be carried out for all humans despite economic stature or scientific development. Article I proclaims that there will be free access to all areas of celestial bodies, and that all activities in space shall be carried out with international peace and cooperation.

The first three articles of the OST prevent any nation from claiming land on the Moon. Because of preservation methods on Earth, the inability to own land makes preservation of the Apollo Landing Sites extremely complicated. A state's ability to preserve a culturally significant site stems from the fact that the site is within its territory and can be protected and managed by the state. The Apollo sites present a difficult legal situation because they are on land that has been internationally agreed upon not to be controlled by any state, and the preservation and management of the site would prove to be extremely difficult. The most realistic way the site could be properly protected is through an international preservation agreement, such as an amendment to the World Heritage Convention.

In 1972, the UN Educational, Scientific, and Cultural Organization (UNESCO) created the World Heritage Convention to help establish criteria for saving the worlds' natural and cultural heritage. The 37-article convention was created to help protect those places essential to understanding human history and without which the entire planet is diminished. Since the creation of this convention, UNESCO has helped protect over 900 natural and cultural places for future generations to experience (United Nations 1972).

Inclusion onto the World Heritage list is decided by a list of ten criteria. There are six criteria for cultural sites and four for natural sites. The Apollo landing sites are eligible for inclusion to the World Heritage List under criteria (i) to represent a masterpiece of human creative genius, (ii) to exhibit an important interchange of human values, over a span of time or within a cultural area of the world, on developments in architecture or technology, monumental arts, town-planning or landscape design, and (iii) to be an outstanding example of a type of building, architectural or technological ensemble or landscape which illustrates (a) significant stage(s) in human history.

The Apollo Landing Sites are achievements that are unparalleled in human history. Only 21 human beings in the history of our species have embarked upon the 239,000-mile journey safely undertaken by the crew of Apollo 11. 600 million television viewers tuned into watch or listen to the Apollo 11 lunar landing alone (Telegraph 2009). The sites can be seen as significant to human civilization as the Pyramids of Giza and the Great Wall of China, but those two World Heritage Sites are on land governed by the laws of a sovereign state, and the Apollo Landing Sites are on territory administered by international treaty.

The World Heritage Convention repeatedly makes the point in its articles that a property has to be on land that belongs to a state in order to be listed. Section II, Article 4 of the World Heritage Convention states: Each State Party to this Convention recognizes that the duty of ensuring the identification, protection, conservation, presentation and transmission to future generations of the cultural and natural heritage referred to in Articles 1 and 2 and *situated on its territory, belongs primarily to that State* [author emphasis]. It will do all it can to this end, to the utmost of its own resources and, where appropriate, with any international assistance and co-operation, in particular, financial, artistic, scientific and technical, which it may be able to obtain (UNESCO 1972).

The language of this article could be used by UNESCO to refuse inclusion of the Apollo Lunar Landing Sites to the World Heritage List because it is not situated on its host states' territory. To make the situation even more muddled, the *Vienna Convention on the Laws of Treaties* states that if two treaties contradict each other, like the WHC and the OST do, and all of the concerned members are party to both treaties, than the earlier treaty applies only to the extent that its provisions are compatible to the newer treaty (United Nations 1967). Because these documents were not designed to work in conjunction with each other than it could be argued that in regards to preservation the OST only denies the ability to protect sites to the extent that the WHC convention allows it to. If UNESCO were to not allow a culturally significant site to be listed because of this reason it would be very hypocritical.

In 1981, "Jerusalem and its Walls" was inscribed on the World Heritage List because of its significance as a holy city of Christianity, Islam, and Judaism, along with having 220 religious monuments (World Heritage 1992a). The nomination of the

World Heritage Site by The Hashemite Kingdom of Jordan was highly controversial among members of the World Heritage Committee at the time, because of the legal situation surrounding control of Jerusalem. In 1947, UN General Assembly Resolution 181 created the State of Israel, but did not include Jerusalem in the state because of a tense political situation. To this day it is governed by a UN Special Committee and is still not legally included into Israel, despite 30 years of continuous occupation. The reasoning behind the controversy creates parallels between "Jerusalem and its Walls," and any potential nomination of the Apollo Lunar Landing Sites.

To make the matter even more complicated the 106 plus objects on the lunar surface at the Apollo 11 Lunar Landing Site are still owned by the United States Government, so it would fall to the US to sponsor this World Heritage Site (NASA 2001).

Articles I, II, and III of the OST declare the lunar surface as an area of international commons and there are parallels that can be made between it and international waters, the sea floor, and the continent of Antarctica. Preservation of cultural resources in areas of international commons is not unprecedented. The Antarctic Heritage Trust was created to help preserve the landmarks from the Heroic Age of Antarctic Exploration Antarctic Heritage Trust (2011a, b).

Antarctica is the only continent left that still contains the first buildings built there by humans. The trust contains two chapters, one based in New Zealand, created in 1987, and one based in Great Britain, created in 1993. The trust's goal is to complete the Ross Sea Heritage Restoration Project, which contains structures from *Southern Cross* (1898–1900), *Discovery* (1901–1904), *Nimrod* (1907–1909), and *Terra Nova* (1910–1913) expeditions.

Because of Articles I, II, and III of the OST and Articles 4, 5, 6, and 11 of the World Heritage Convention, preserving the Apollo Lunar Landing Sites may not be possible by traditional methods. Because of articles 30–32 of the *Vienna Convention on the Laws of Treaties*, the interaction between the OST and the WHC creates a complicated legal situation. The analysis of the four treaties and conventions written after the OST will support this statement. It is clear from the information presented in this section that when COPUOS drafted the OST, they were only concerned with the immediate matters at hand. Because the treaty was created before any human had visited the lunar surface, or the creation of the World Heritage Convention, preserving human artifacts was not a pertinent issue. An amendment to the World Heritage Convention, or a new form of treaty or convention would be beneficial, but not necessary for proper legal protection of these sites. Despite the amount of information cited that supports this theory, there are instances where legal protection for objects and sites has been enacted despite the language of the law.

Of the other four treaties created to govern human usage of the heavens only one of them plays any sort of role in this argument. The 1968 *Agreement on the Rescue of Astronauts, the Return of Astronauts, and the Return of Objects Launched into Outer Space*, commonly referred to as the 1968 Rescue Agreement, followed the OST shortly after its creation but deals specifically with aiding any astronaut that finds itself in danger outside of our atmosphere (Beckman 2003).

The Convention on International Liability for Damage Caused by Space Objects, created in 1972 and commonly known as The Liability Convention was created because of the frequently asked questions: "what happens when something goes wrong in space? And who pays for it when it does?"

The *Liability Convention* is largely an expansion on the principles of liability for damage from outer space objects that was introduced in Articles III and VII of the 1967 Outer Space Treaty. It contains 28 Articles that address issues ranging from determining the launching state when multiple parties are involved, determining liability for destruction caused by space debris, and who is liable when a private party launches an object into space.

The fourth treaty drafted by COPUOS to govern human usage of outer space is the 1976 *Convention on the Registration of Objects into Outer Space,* commonly referred to as the 1976 *Registration Convention.* The *Registration Convention* was created for the straightforward reason of registering of spacecraft so that should an event occur where there was damage or loss of life, the spacecraft that caused the event could be identified.

As previously discussed the Liability Convention was created to answer the question of "What happens when something goes wrong in space? And who pays for it?" The *Registration Convention* was created to make sure the *Liability Convention* worked properly.

From 1967 to 1979, COPUOS created five treaties and conventions that began the process of creating guidelines for human behavior on other celestial bodies, and in outer space. The fifth and final treaty created during the initial phase of drafting laws is the *Agreement Governing the Activities of States on the Moon and Other Celestial Bodies* (1979). This legislative document, commonly referred to as the *Moon Agreement,* is far and away the most controversial of any of the laws governing the heavens. Of the five space laws created by COPUOS, the *Moon Agreement* is the only one not to be signed by the US, the Russian Federation, the People's Republic of China, or Great Britain. Despite these nations not signing the treaty, it still collected enough signatures to be entered into force, and is considered international law.

Much like the laws before it, the *Moon Agreement* was created to establish guidelines for the human usage of resources on other celestial bodies, but specifically on our closest celestial body. The major points of the treaty are to safely develop and manage lunar resources, to create more opportunities to use these resources, and to share whatever benefits come from those resources. Like the Registration Convention, Liability Convention, and Rescue Agreement, the *Moon Agreement* was derived from articles in the OST and then expanded upon. Unlike the OST, whose language was written by US and Soviet delegates, the language of the Moon Agreement was written largely by other nations (Cristol 1980).

The scope of the two documents is very similar in regards to what they were created to accomplish. Both the OST and the *Moon Agreement* call for the peaceful usage of outer space, and the moon. Both treaties call for international cooperation, especially when working on scientific research or exploration. Both treaties call for international responsibility for a nation's activities, and freedom from interference from another nation's activities in outer space.

Neither treaty allows for national appropriation of land on the lunar surface. The *Moon Agreement* takes that one step further and does not allow for any personal or corporate appropriation of land on the lunar surface, or the surface of any celestial body unless it is administered by an international body. Because prior laws had not specifically forbade private appropriation of land on the lunar surface, entrepreneurs saw an opportunity to create businesses selling land on the moon to gullible customers. Dennis Hope is one example of an entrepreneur who has been illegally selling property on the Moon, Mercury, Venus, Mars, and the moon of Jupiter, Io (Lunar Embassy 1980).

While the similarities between the *Moon Agreement* and the OST are numerous, the differences between the two really define this treaty. The *Moon Agreement* is a highly controversial legal document that has been argued by legal scholars and space enthusiasts since its creation. Because there are articles in the treaty that are relevant to preservation of the Apollo Lunar Landing Sites the *Moon Agreement* treaty must be discussed.

The *Moon Agreement* introduces the concept of the "Common Heritage of Mankind," to international space law in Article XI of the treaty. This article is considered to be the only reason this treaty was drafted, because it is one of the few articles present in the treaty that is not present in any of the other treaties (Gangale 2009). If an area has been labeled "Common Heritage of Mankind" (CHM) it means that no one can legally own that area, though everyone manages the area, claims of national sovereignty do not exist. This means that while no one state or group of states can claim to own an area such as the lunar surface or international waters, as human beings were are all responsible for the care of it (Harminderpal 1995). While there is no universally accepted definition of CHM areas there are five general elements.

1. The CHM area is not subject to appropriation
2. All states share in the areas resource management
3. States must share the benefits derived from exploitation of area resources
4. The CHM area must be dedicated to peaceful purposes exclusively
5. The CHM area must be preserved for posterity (Harminderpal 1995)

When the CHM principle is implemented it usually involves the creation of income sharing schemes from the natural resources extracted from the area. Developing nations view the CHM principle as a way to level the playing field between themselves and developed nations. In regards to the lunar surface, developing nations viewed this as a way to have one source extract the materials, and then distribute them properly among all of the involved nations. Developed nations disagree with this notion of CHM because it would alter the current structure of economic power. The developed nations believe that they should be allowed keep any profits earned from their ability to access areas that other nations cannot. This is considered to be one of the major reasons that the major space powers have not as of yet signed the *Moon Agreement*.

It has also been argued that the *Moon Agreement* was not signed by the space powers because of the time period in which it was written (Harminderpal 1995).

When the call for a new treaty had been proposed, the US was in the middle of its manned lunar landing program; by the time it was written no human being had been on the moon for seven years. The US abandoned its manned lunar landing program with the completion of Apollo 17 in 1972, and the Soviets had all but given up on the task by 1974. By the time Neal Armstrong had landed in Tranquility Base, the US government already had plans for a manned mission to Mars, but those plans were scrapped upon the completion of the Apollo program. When the agreement was in its infancy in the early 1970s human settlement of the moon seemed just around the corner; by the time it was completed in 1979, lunar settlement was decades away.

Unlike its predecessors, the *Moon Agreement* does have one specific article that deals with the preservation on the lunar surface. It could be argued that not only would this article allow for the preservation of the human artifacts left on the lunar surface, but the most important feature left on the moon Neal Armstrong's footprint. Article 7 Paragraph 3 of The Moon Agreement states the following: "States Parties shall report to other States Parties and to the Secretary-General concerning areas of the Moon having special scientific interest in order that without prejudice to the rights of other States Parties, consideration may be given to the designation of such areas as international scientific preserves for which special protective arrangements are to be agreed upon in consultation with the competent bodies of the United Nations" (UNESCO 1979).

While it could be argued that archeology and preservation are not "hard" sciences in the same category as astrophysics and molecular biology, the area in question is of unquestioned importance to human history. It is clear that when Article 7 was written it was created to safeguard areas of the moon that contain marketable minerals (Popular Mechanics 2004), or the water that is trapped in the moon's poles (CNN 2012), and not the array of equipment left on the lunar surface by astronauts. Furthermore, the view of Tranquility Base and the lunar land sites as historic-era *archaeological* sites is well-established within the historic preservation community, and the use of the scientific method in the professional of archaeology place the field—and sites—within the reasonable bounds of "science."

Protecting the Apollo 11 Lunar Landing Site Through United States Preservation Law

The artifacts left on the lunar surface by the Apollo astronauts all have one thing in common. They were brought to the Moon by Americans, and are still US government property. Because those objects are still owned by the US government they can be protected by the US government by nominating them as a National Historic Landmark or a National Monument. It is completely legal for the US to do so; hence an analysis of US preservation law is necessary.

In 1966, the National Historic Preservation Act (NHPA) was passed, which created one of the first comprehensive historic preservation laws in the nation

(National Historic Preservation Act). The NHPA established State Historic Preservation Officers, the National Register of Historic Places (NRHP), and National Historic Landmarks (NHL). The creation of this law established a mechanism for federal agency decision making regarding private and public projects that have the potential to adversely affect significant sites, buildings, structures or objects in American history. Pertinent to the growing private sector space exploration industry is the fact that the NHPA requires that federal agencies who will provide federal funding, approval, or permits for a private sector project or activity, must take into account whether or not that undertaking will damage or destroy significant historic properties.

Sites that have been determined to be eligible for listing as National Historic Landmarks are the most significant places, sites, buildings, and objects in American history. In contrast, the NRHP are significant on national, state and local levels, but are not National Historic Landmarks. In order to be included as a National Historic Landmark the site must be eligible under one of the six following criteria:

1. That are associated with events that have made a significant contribution to, and are identified with, or that outstandingly represent, the broad national patterns of United States history and from which an understanding and appreciation of those patterns may be gained;
2. That are associated importantly with the lives of persons nationally significant in the history of the United States; or
3. That represent some great idea or ideal of the American people; or
4. That embody the distinguishing characteristics of an architectural type specimen exceptionally valuable for a study of a period, style or method of construction, or that represent a significant, distinctive and exceptional entity whose components may lack individual distinction; or
5. That are composed of integral parts of the environment not sufficiently significant by reason of historical association or artistic merit to warrant individual recognition but collectively compose an entity of exceptional historical or artistic significance, or outstandingly commemorate or illustrate a way of life or culture; or
6. That have yielded or may be likely to yield information of major scientific importance by revealing new cultures, or by shedding light upon periods of occupation over large areas of the United States. Such sites are those which have yielded, or which may reasonably be expected to yield, data affecting theories, concepts and ideas to a major degree (National Historic Preservation Act 1966).

It can be argued that the first human footsteps on the Moon, or any celestial body meet National Historic Landmark Criterion 1. The culmination of the Apollo Program was made possible by decades of research into physics, aerospace engineering, and chemistry.

The technology used in the creation of the Apollo 11 rockets and lunar modules was remarkably advanced for its time, so much so that forty-five-plus years later humans now lack the technology to return to the lunar surface. What remains at the Sea of Tranquility could be considered the height of human technology for the mid-20th century. The landing craft present at all Apollo Lunar Landing Sites present the only technology ever created that is capable of sustaining human life on the lunar surface, all other objects on the moon were created for unmanned missions. The objects on the moon more than qualify for inclusion as a National Historic Landmark under Criteria 4 and 5 because the majority of the objects left at Tranquility Base only exist on Earth as prototypes.

The site is also eligible under Criterion 3. However, the idea of humans being on the moon is not new. An American did not create the idea, nor is the idea a strictly American ideal. However, getting to the moon was done in a very American fashion. It started when President Kennedy declared: "This nation should commit itself to achieving the goal, before the decade is out, of landing a man on the moon and returning him safely to Earth" (NASA History Office n.d.). Kennedy knew very well that the United States did not possess the technology to do so at the time of his speech to the U.S. Congress. That did not deter him from making the race to the moon one of the major priorities of America in the 1960s. Nine years, countless man-hours, and billions of government dollars later, an American citizen became the first member of his species to set foot on another celestial body (Butowsky 1984).

Because these objects are still United States government property, they can be considered for inclusion as a National Historic Landmark (Gibson 2001). But because the moon cannot be owned by anyone, only the objects—and not the site—can be nominated for inclusion under US law. The states of California and New Mexico have listed the objects left on the lunar surface by Apollo 11 Astronauts in their state historic registers because of the connections the states have to producing or testing those objects (Westwood et al. 2010; O'Leary et al. 2010).

In 1984, the National Park Service conducted a theme study of all of the sites associated with the US space program from its infancy to its successful landing of the first man on the moon and beyond (Butowsky 1984). This study advised the US government to include 24 sites as National Historic Landmarks that were associated with significant achievements in aeronautics and the history of our space program. Of those 24 sites, six of them are directly related to Apollo missions that landed the first humans on the moon. Prior to this study, Cape Canaveral Air Force Station was listed as a National Historic Landmark on April 16, 1984 (NPS n.d.). Launch Complex 39 at the Kennedy Space Center, from where all of the Apollo missions lifted off, was listed as a National Historic Landmark on May 24, 1973 (Butowsky 1984).

It is ironic that all of the buildings, facilities, test sites, rockets, test modules, and equipment (i.e. objects) that led up to one of the most significant moments in human history are protected under US law, yet the objects that are associated with the pinnacle achievement of the same "historic district of objects" are not because they are located on the lunar surface.

The inclusion of the objects at the Apollo Lunar Landing Sites as a National Historic Landmark is complicated because it is not situated on United States land. However, the idea of a National Historic Landmark not on US territory is not foreign. There are several National Historic Landmark listings that are beyond US soil, or on US protectorates. The World War II battlefields on the Pacific Island nations of Palau, and Micronesia were listed as National Historic Landmarks in 1985 (NPS n.d.).

The battlefields on Palau in Micronesia are now located on the land of sovereign nations, but at the time of their creation these nations were trust territories under the control of the United States (Gibson 2001). The American Legation in Morocco is another National Historic Landmark located outside of US borders (Gibson 2001). However, it is located on the site of a former embassy so it is on US government land (Gibson 2001). These three sites do not provide as strong of a precedent needed for a complex legal matter like the one presented by the Apollo Lunar Landing Sites, but they do indicate that National Historic Landmarks can be located outside of US territory.

The National Historic Landmark that presents an appropriate precedent for the Apollo Lunar Landing Sites is "Old Ironsides." The *USS Constitution* is a US Navy frigate that earned its nickname for performing gallantly during the War of 1812. The ship is a reproduction of its original self, and was nominated as a National Historic Landmark in 1960 (NPS n.d). The *USS Constitution* is similar to the previously mentioned National Historic Landmark sites because there are times where it is not located inside of US territory, and is mobile, like the Eagle lander at Tranquility Base.

Those sites, despite not being on US land are on land owned by a sovereign nation. The *USS Constitution* is different because it is the only National Historic Landmark that can move into international waters under its own power, which are an area of international commons. The *USS Constitution* would be a National Historic Landmark whether it is docked at a US port, a Russian port, or sailing on the high seas (Gibson 2001).

The precedent set by the inclusion of the *USS Constitution* as a National Historic Landmark is one that could be followed in attempts to preserve the objects at the Apollo Lunar Landing Sites, especially in conjunction with the number of Apollo testing objects and sites listed already. Because the *USS Constitution* has the ability to sail in international waters, or in the territorial waters of a sovereign nation other than the one whose register it is on creates an interesting legal precedent. One of the major fears when nominating the objects at the Apollo Lunar Landing Sites is other countries thinking that the US is claiming sovereignty over the lunar surface by protecting the objects left there. The case of the *USS Constitution* shows how an object legally protected by a sovereign state can exist in an area of international commons.

What is needed to further the argument towards protection is proof that an object can be protected in an area that belongs to no one, without claims of sovereignty. Because of the 1980 amendments to the NHPA, the *USS Constitution* is eligible for nomination to the WHL and would provide the perfect precedent for the objects at the Apollo Lunar Landing Sites. Because there is a National Historic Landmark that exists in an international area of commons, it is important to look at a similar World Heritage Site.

The Old City of Jerusalem and Its Walls

The City of Jerusalem has been a holy city for Christianity, Islam, and Judaism for 3,000 years. The city contains over 220 religious monuments and statues for those seeking relics of their respective religion. Along with its religious significance, the city also exists in a state of legal significance that creates similarities between itself and the lunar surface. The matters discussed here are controversial legal issues stemming from thousands of years of Middle Eastern regional politics. This chapter does not claim to understand the political situations in the Middle East, nor does it take a side in the Palestinian or Israeli conflict to govern the city of Jerusalem, but rather uses the legal precedent established by the declaration of Jerusalem as *corpus separatum* to further the argument for preservation of the Apollo Lunar Landing Sites by inclusion to the World Heritage List.

In 1947, when the United Nations passed General Assembly Resolution 181, creating an Israeli State, they declared the city of Jerusalem "shall be established as *corpus separatum* under a special international regime administered by the United Nations" (Avalon Project 1996). The term *corpus separatum* is Latin for "separated body." In regards to governing the city of Jerusalem, the UN wanted a "special international regime and shall be administered by the United Nations" (Avalon Project 1996). The UN did not want the city to be considered part of Israel, Palestine, or Jordan because it did not want to support the claims of any state to control Jerusalem.

When the UN established the state of Israel they understood the complicated situation they were creating in the Middle East. Because the religions of Christianity, Islam, and Judaism all consider Jerusalem to be a Holy City, the UN realized the international importance of the city and its need for preservation. If the city of Jerusalem were given outright to the new state of Israel, knowing that Palestinians considered the city their capitol, it would have increased the amount of friction in the area. The decision to establish an international regime to govern Jerusalem, and its holy sites and monuments instead of granting sovereign rule of the city to Israeli, Palestinian, or Jordanian leaders was furthered along by the passing of UN General Assembly Resolution 303 (United Nations General Assembly 2012). The situation in Jerusalem is made even more complicated by the fact that Israel has occupied West Jerusalem since 1948 and East Jerusalem since the end of the 6 Day War in 1967. Despite 30 years of occupying the entire city, the majority of the world does not recognize Israeli sovereignty over the city of Jerusalem (Quigley 1996).

These General Assembly decisions establish a set of legal parameters that allow the city of Jerusalem to exist on land not governed by any state. Legally speaking, there is little difference between the land Jerusalem is situated on, and the Sea of Tranquility. The laws that govern these two internationally significant sites were created by the United Nations, and they are not controlled by any sovereign state. The city of Jerusalem and its complicated legal situation are important to the preservation of the Apollo Lunar Landing Sites because of the city's inclusion to the World Heritage List in 1981.

"Old City of Jerusalem and its Walls," was proposed for inclusion into the World Heritage List because of its religious significance to three of the world's major religions, its cultural significance because of the vast amounts of religious structures and monuments, and its importance as one of the oldest cities on the planet. The proposal of this site to be included to the WHL created controversy because the country of Jordan proposed the site. While Jordan is a geographical neighbor of the city, it does not have a sovereign claim to the city. The proposal of a site by a state that the site is not situated in directly violates Article 11, paragraph 3, of the World Heritage Convention (United Nations 1972).

In documents detailing the discussions of the committee members, there is an overwhelming support shown by all in attendance for the high level of significance for the site, and for Jordan's competence in administering preservation to the city. Where the delegations found fault was Jordan's legal ability to preserve the site. A letter to the committee from the Jordanian representative outlines the reasoning behind Jordan's sponsoring the site for its religious significance, and the deterioration of its heritage. That letter specifically states that this nomination was in no way a claim of sovereignty to the city of Jerusalem by the Hashemite Kingdom of Jordan (UNESCO 1981).

A final vote to establish the site passed 14 to 1, with 5 abstentions, and several of the delegates felt the need to voice their concerns for the site before tallying their vote. The delegations from Australia, France, Germany, Italy, and Switzerland explained their abstention from the vote because UN General Assembly Resolution 181 established a situation where Jerusalem is not a part of a sovereign nation, hence Jordan as a sovereign power, has no right to propose its preservation.

For the first time in hundreds of years, European powers felt that not meddling in Middle Eastern politics was the best option. The delegations from Argentina, Cyprus, Egypt, Nepal, and Zaire all explained their endorsement for including the site as being based off of the cultural significance of the site, and that the proposal does not impose any claim of sovereignty over the city. The delegation from the United States was the sole vote against the inclusion of the site because it went against Article 11, Paragraph 3, of the World Heritage Convention.

The addition of "Old City of Jerusalem and its Walls" to the World Heritage List should be referenced in any attempt to include the Apollo Lunar Landing Sites to the list. The legal similarities between Jerusalem and the lunar surface are enough to create a discussion about the two sites. With multiple nations voting to include this site, despite it not being located on the territory of the sponsor, creates legal precedent that a site like the Apollo Lunar Landing Sites would need for consideration. The fact that nations which voted for its inclusion made remarks pertaining to the sponsorship of the site in no way being viewed as a claim of sovereignty also helps the argument.

However, Jordan was successful in its sponsorship because Jerusalem is being threatened by urban construction projects and mass deterioration of its monuments and structures, and something needed to be done to stop it (World Heritage List 1992b). In order for any sponsoring state to be successful it would need to properly present the dangers at the Apollo Lunar Landing Sites, and the proper legal background to show that the idea is not as farfetched as it may seem.

The international conventions and treaties created to govern human usage of the heavens were created for various reasons. They were drafted to stop developed nations from treating celestial bodies the same way 16th century European explorers treated the New World by claiming land for their crowns through various cultural ceremonies. These documents were written to stop the US and the USSR from placing nuclear weapons on the Moon. They were fashioned to help astronauts should they ever find themselves in danger in outer space. The laws were shaped to place liability upon a state that did something the wrong way, and ended up damaging another states property. The laws were drafted to keep track of everything humans sent up into space, and to take care of the immediate needs of space faring states at the time of their drafting. One thing the laws were not created to do is preserve locations of human cultural significance on celestial bodies.

That fact that these treaties did not take preservation into account should not be surprising for several reasons. First, the OST was ratified 2 years before a human presence was made on the lunar surface. The laws at that point were theoretical, and human beings have a tendency to be more reactive than proactive by only preserving sites of historical importance once they become threatened. Because no human had ever been on the moon before, how could it be threatened?

More importantly, because no one had been there, how could anyone protect it? Human occupation of the moon took place intermittently during a brief three-year window from 1969 to 1972, and only twelve humans have ever set foot upon the lunar surface. That means the only significant place on our planet that fewer people have been to than the moon is the bottom of the Mariana Trench, the lowest point on the planet.

Arguing that these laws do not allow for the preservation of archeological sites on the lunar surface is complex. Just because that idea was not directly factored into the drafting of this legislation does not mean that it is impossible. These conventions and treaties consistently ask for international cooperation in matters pertaining to human usage of the heavens, and that is exactly what is necessary for the Apollo Landing Sites to be protected by the World Heritage Convention.

The lag time of historic preservation considerations in space exploration follows similar concerns on Earth. Human beings have been using the oceans for trade, travel, and warfare for thousands of years. There are an unimaginable number of shipwrecks and culturally significant objects lying on the ocean floor, and it was only in 2001 that UNESCO created the *Convention on the Protection of Underwater Cultural Heritage* to safeguard those objects (UNESCO 2001). Antarctic exploration reached its peak during the turn of the 20th century, when Roald Amundsen reached the South Pole in 1911. The Antarctic Heritage Trust was created in 1987; 76 years after the first humans reached the South Pole.

Historic preservation takes both time and perspective, but more importantly, a proactive stance. Enough time has passed to preserve the Apollo Lunar Landing Sites for their significance to be appreciated, but human beings need far more perspective on the situation to truly appreciate how important Neil Armstrong's footsteps are to the history of our human species.

Conclusions

Current international laws were created with preservation in mind. It was, how-ever, the preservation of human lives from a nuclear threat, not the preservation of human archeological sites that were of paramount concern. The language of the OST may prevent any nation from protecting the Apollo Lunar Landing Sites as a whole (meaning both the landing site and the objects Armstrong and Aldrin left behind). Articles 1, 2, and 3 strictly forbid the sovereign claim of land on the Moon, and other celestial bodies, but there is no language that strictly forbids the protection of the objects at the sites. The lack of preventative language stems from the absence of a human presence on the Moon at the time of the drafting of the OST. This creates a legal gray area in a largely theoretical area of law.

The analysis and interpretation of current international law did not present a clear answer to the question because there is not one, at least not in regards to international preservation. This is a highly complicated legal matter that deals with the relationship between multiple international conventions and treaties covering multiple topics across more than half a century. Some of these treaties were not created to work in conjunc-tion with each other, but it is not hard to imagine that it will be necessary to weave them into a more coherent treaty in the future. Amending the OST or World Heritage Convention is the most viable option for protecting the sites on an international level.

In regards to national protection, the objects resting on the lunar surface at each Apollo Lunar Landing Site are still claimed by the United States government. Because the US government has not abandoned the materials, it can legally pre-serve and protect them. Doing so is another story. It is within Presidential power to protect the objects at the sites via Executive Order, through creation of a National Monument, or either the President or the Senate could approve sites as a National Historic Landmarks. The US government could take such action at any time.

The question of whether or not individual citizens of the United States *should* protect the Apollo Lunar Landing Sites is an entirely different pursuit. Should the United States ignore legal gray areas and protect the sites because of their signifi-cance as sites of human historical and scientific achievement in the same way that the Hashemite Kingdom of Jordan ignored the World Heritage Convention and protected "Jerusalem and Its Walls" for its significance as a historical and religious site? The answer to that question is yes, the United States should preserve these sites as National Historic Landmarks. Here is why.

The Apollo landing sites are, on one hand, a reflection of American technologi-cal and scientific prowess. They are manifestations of the design, fabrication and operation of the most complicated machines humanity has devised. The sites are worthy of protection as representations of those achievements. But there are other reasons the sites deserve protection.

These sites should be included as National Historic Landmarks because they represent what may, in historical retrospect, be the pinnacle of American bravado. The effects the space race had on our culture cannot be measured in any one way. The technological advances brought on by the audacity of President Kennedy's claim radically altered the economic and social fabric of this country for decades after. By launching America into a lunar crusade, Kennedy created a culture of innovation and discovery just as vital to this country's history as any of the other social movements that have taken place during the 20th century.

The United States landed twelve men on the Moon. That feat is, a half century later, unmatched. A sense of pride came from Americans changing space from a fictional setting of novels and television shows to a real place. The future was tangible; it was right around the corner. That future was the creation of the culture of innovation that was fueled by the hope of discovery. The Apollo Lunar Landing Sites are a physical manifestation of that innovation, hope and discovery. That is why the US *should* preserve these sites.

The protection of these lunar sites would become a watershed moment for preservation. The preservation of these extraterrestrial sites would mark the changing of a thought process that has enveloped the field of preservation since its inception. Preservation as a field needs to adapt to this or the profession will become irrelevant. The preservation of a site that was the launching point of the future could become a relatively uncomfortable task for a profession that is rooted primarily in domestic and commercial architecture and embraces technological and industrial sites timidly. Preserving this site would shift, dramatically, what we define as worthy of preservation and protection.

There are, to be sure, reasonable arguments for not protecting this site. Primarily, the fear that stems from causing an international incident is one of the largest reasons. The protection of this site could be viewed as a direct violation of the first three articles of the OST (Milstein 2008). This view is not correct, because the designation of a National Historic Landmark can be a purely symbolic designation meant to remind people of the greatness their country and its citizens have accomplished. The designations are not in any way a claim of sovereignty, nor would it mean the US would go to war with any nation or corporation that disturbed or damaged the sites. Designating these sites for protection is nothing more than a national pride building celebration.

Sadly, even protection on a national level does not completely solve the issue. To call this only a significant moment in US history is shortsighted. This is a moment in *human* history. The plaque left on the Moon by Aldrin and Armstrong called themselves "ambassadors of the planet Earth," not just ambassadors from the US. Like the space race itself, protection of the site should be conducted on a national level by the US first. Then, once we have accomplished that mission, it will be more palatable for others nation to help protect the sites.

References

Antarctic Heritage Trust. (2011a). *Welcome to the Antarctic Heritage Trust*. Accessed February 9, 2012. www.nzaht.org

Antarctic Heritage Trust. (2011b). *The expedition bases*. Accessed February 9, 2012. www.nzaht.org/aht/theexpeditionbases

Beckman, R. C. (2003). 1968 Rescue agreement: An overview. Paper presented at the United Nations/Republic of Korea workshop on space law, Daejeon, Republic of Korea, November 3–6, 2003

Butowsky, H. (1984). *Man in space—A national historic landmark theme study*. Washington DC: US National Park Service, US Department of Interior. Accessed February 28, 2012. http://historicproperties.arc.nasa.gov/downloads/man_in_space_butowsky.pdf.

CNN. (2012). *NASA finds significant water on moon*. Accessed February 5th, 2012. http://articles.cnn.com/2009-11-13/tech/water.moon.nasa_1_lunar-crater-observation-anthony-colaprete-solar-system?_s=PM:TECH

Cristol, C. Q. (1980). International liability for damage caused by space objects. *The American Journal of International Law, 74*, 342–369.

Dembling, P. G., & Arons, D. M. (1967). The evolution of the outer space treaty. *The Journal of Air Law and Commerce, 33*, 421.

Gangale, T. (2009). *The development of outer space: Sovereignty and property rights in international space law*. Santa Barbara: ABC-CLIO.

Gibson, R. (2001). Lunar archeology: The application of federal historic preservation law to the site where humans first set foot upon the moon (Department of Anthropology, New Mexico State University). *Masters Thesis*.

Harminderpal, S. R. (1994–1995). The common heritage of mankind and the final frontier: A revaluation of values constituting the international legal regime for outer space activities. *The Rutgers Journal of Law, 26*, 229.

Hosenball, S. N. (1979). The United Nations Committee on the peaceful uses of outer space: Past accomplishments and future challenges. *Journal of Space Law, 7*, 95–115.

The Avalon Project. (1996). *The United Nations general assembly resolution 181*. Accessed February 20, 2012. www.yale.edu/lawweb/avalon/un/res181.htm.

The Lunar Embassy. (1980). *Extraterrestrial property and space law: Fact and fiction*. Accessed January 30, 2012. http://www.lunarembassy.com/lunar/shops.lasso?-database=aa654s5677556pr&-layout=US$_pr9981_en&-response=index_e.lasso&-NoResultsError=index_e.lasso&-token.affindex=&-token.trackindex=2128702&-token.rn=91966635&-token.cs=US$&-token.rs29=33&-token.rscd=LE&-token.firstlogin=&-token.skip=&-show

The Miller Center. (2012). *Lyndon B Johnson—May 1966*. University of Virginia. Accessed January 23, 2012. http://millercenter.org/scripps/archive/presidentialrecordings/johnson/1966/05_1966

Milstein, M. (2008). *NASA looks to protect historic sites on the moon*. Accessed February 19, 2012. http://www.smithsonianmag.com/science-nature/nasa-looks-to-protect-historic-sites-on-the-moon-47186092/

NASA. (2001). *Recommendations to space faring entities: How to protect and preserve the historic and scientific value of U.S. government Lunar Artifacts*.

NASA History. (n.d). *The decision to go to the moon: President John F Kennedy's May 25, 1961 speech to congress*. Accessed February 18, 2012. http://history.nasa.gov/moondec.html.

National Park Service. (1966). *The national historic preservation Act of 1966*. National Park Service: Washington DC.

National Park Service. (1966a). Summary of the national historic landmarks criteria for evaluation. *National Register Bulletin 15*. Accessed February 28, 2012. http://www.nps.gov/nr/publications/bulletins/nrb15/nrb15_9.htm

National Park Service. (n.d.) *Listing of national historic landmarks by state*. Accessed February 16, 2012. www.nps.gov/nhl/designations/lists/micro01.pdf

National Park Service. (n.d.) *National historic landmark program: Cape Canaveral air force station*. Accessed February 28, 2012. http://tps.cr.nps.gov/nhl/detail.cfm?ResourceId=1885&ResourceType=District

National Park Service. *Maritime landmarks: Large vessels*. Accessed February 16, 2012. www.nps.gov/maritime/nhl/lvesnhl.html

O'Leary, B., Bliss, S., Debry, R., Gibson, R., Punke, M., Sam, D., Slocum, R., & Vela, J. (2010). The artifacts and structures at tranquility base. Nomination to the New Mexico state register of cultural properties. Accepted by unanimous vote by the New Mexico cultural properties review committee on April 10, 2010.

Popular Mechanics. (2004). *Mining the Moon.* Accessed February 5, 2012. http://www.popularmechanics. com/science/space/moon-mars/1283056

Quigley, J. (1996). Sovereignty in Jerusalem. *Catholic University Law Review, 45,* 765.

Telegraph. (2009). *Apollo 11 moon landing: Ten facts about Armstrong, Aldrin, Collins' Mission.* Accessed February 9, 2012. www.telegraph.co.uk/science/space/5852237/Apollo-11-Moon-Landing-ten-facts-about- armstrong-and-collins-mission.html

United Nations. (2001). Committee on the peaceful use of outer space. *United Nations committee on the peaceful uses of outer space members.* Accessed February 21, 2012. www.oosa.unvien na.org/oosa/COPUOS/members.html.

United Nations. (1972). *The world heritage convention.* Geneva: UN.

United Nations. (2001). *The convention on protecting underwater cultural heritage.* Geneva: UN.

United Nations. *The world heritage list.* Accessed February 9, 2012. http://whc.unesco.org/en/list

United Nations. (1981). World heritage committee: 1st extraordinary session, Paris 1981. Accessed February 14, 2012. http://whc.unesco.org/archive/repext81.htm#annex4

United Nations. (2012). *General assembly resolution 303.* Accessed February 20, 2012. http://w ww.jewishvirtuallibrary.org/jsource/UN/unga303.html

United Nations. (1967). *Treaty on the principles governing the activities of states in the exploration and use of outer space, including the moon and other celestial bodies.* Geneva: United Nations Committee on the Peaceful Uses of Outer Space.

United Nations. (1968). *The agreement on the rescue of astronauts, the return of astronauts and the return of objects launched into outer space.* Geneva: United Nations Committee on the Peaceful Uses of Outer Space.

United Nations. (1972). *The convention on international liability for damage caused by space objects.* Geneva: United Nations Committee on the Peaceful Uses of Outer Space.

United Nations. (1976). *The convention on the registration of objects into outer space.* Geneva: United Nations Committee on the Peaceful Uses of Outer Space.

United Nations. (1979). *The agreement governing the activities of states on the moon and other celestial bodies.* Geneva: UN.

United Nations. (2006). *Treaty signatures.* Accessed January 23, 2012. http://www.oosa.unvienna .org/oosatdb/showTreatySignatures.do?d-8032343-p=3&statusCode=PARTY&treatyCode= RA&stateOrganizationCode=

Westwood, L ., & Gibson, R., O'Leary, B., & Versluis, J. (2010). Nomination of objects associated with tranquility base to the California state historical resources commission. Accepted by unanimous vote to the California Register of Historical Resources, January 29, 2010. Accessed February 16, 2012. www.parks.ca.gov/pages/1067/files/tranquility%20base_draft.pdf.

World Heritage List. (1992a). *World heritage list.* Accessed February 11, 2012. http://whc.unesco. org/en/list/

World Heritage List. (1992b). *Old city of Jerusalem and its walls.* Accessed February 9, 2012. http://whc.unesco.org/en/list/148/

World Heritage List. (1983). *World heritage committee: Report of the sixth session.* Accessed February 20, 2012. http://whc.unesco.org/archive/repcom82.htm#jerusalem

Chapter 9
Historic Preservation on the Fringe: A Human Lunar Exploration Heritage Cultural Landscape

Lisa D. Westwood

Abstract This chapter presents a systematic overview of international avenues for preserving space heritage. The author examines the patchwork of state, national and international laws, ordinances, executive orders and guidelines that has yet to expand vertically into outer space. It discusses the preservation efforts to seek formal designation in the US of the first lunar landing site at Tranquility Base. Applying a landscape approach to early space heritage, the author explores the possibilities and challenges associated with designating a World Heritage List district of space related sites that spans multiple countries as well as planetary bodies.

Introduction

The disconnect between existing historic preservation frameworks and nontraditional cultural resources is growing. While extant legal frameworks have grown horizontally to address loopholes jeopardizing humanity's heritage, this patchwork of state, national, and international laws, ordinances, executive orders, and guidelines has yet to expand vertically, into the heavens. This became more apparent with efforts to seek formal designation of Tranquility Base, the Apollo 11 lunar landing site on the Moon, which began with listing on historical registries at the state level in the United States.

We now can broaden that view to encompass many other space exploration properties on Earth, on the Moon, and beyond, as efforts reach toward global historic preservation. This chapter applies a cultural landscape approach to early space exploration heritage, and explores the possibility and challenges associated with designating a World Heritage List district of related sites and properties that spans not only multiple countries, but planetary bodies as well.

L.D. Westwood (✉)
California, USA
e-mail: lisa@lisawestwood.com

© Springer International Publishing Switzerland 2015
B.L. O'Leary and P.J. Capelotti (eds.), *Archaeology and Heritage of the Human Movement into Space*, Space and Society, DOI 10.1007/978-3-319-07866-3_9

The Regulatory Framework of Historic Preservation

Since the enactment of the American Antiquities Act in 1906, the United States Congress enacted nearly four dozen laws, regulations, executive orders, and guidelines specific to historic preservation matters that collectively seek to protect and preserve America's history (Table 9.1). This regulatory context has been designed to manage cultural resources in the traditional sense—namely, earth-bound archaeological sites and historic buildings with definable boundaries and clear legal jurisdiction. This two-dimensional regulatory framework has been patched together over the past century to reactively address the growing concern over the loss of our nation's heritage caused by the encroachment of modern human culture into our collective past.

A cursory examination of the scope of these pieces of legislation reflects then—current political trends, events, and worldview. World War II and the Korean War, coupled with the expansion of technology used by news media, effectively shrank the world and raised awareness about the consequences of warfare on the environment and on relics of our history. Accordingly, the mid-1960s saw the enactment of several key pieces of legislation, including the American Battlefield Protection Act of 1966, the National Historic Preservation Act of 1966, and the National Environmental Policy Act of 1969. Similarly, the politically charged American Indian Movement of the 1960s and 1970s for rights of self-determination brought about the passage of American Indian Religious Freedom Act in 1978 and the Native American Graves Protection and Repatriation Act in 1991.

Following this pattern, one might expect that legislation enacted in the 1970s and 1980s—a period of heightened human space exploration via the Apollo program, Space Shuttle program, and construction of the International Space Station and Hubble telescope—might reflect the growing assemblage of space heritage resources. However, none of the laws, guidelines, or executive orders passed between 1960 (the conservative beginning of the "space age") to the present day directly address historic preservation of space heritage.

Moreover, in recent years, there has been an increasing awareness of other types of cultural resources, including cultural landscapes, discontiguous archaeological districts, and moveable objects or structures, like old battleships and early space exploration vehicles. As we move into these "fringe" areas of cultural resources, which represent some highly significant milestones in our heritage, the idea of legislating historic preservation begins to grind uncomfortably against the existing regulatory framework.

The Foundations of Historic Preservation Law in the US

In authorizing the National Historic Preservation Act (NHPA) in 1966—just 3 years prior to Apollo 11—Congress recognized the fact that there are physical manifestations of our Nation's history and culture that are important, in jeopardy, irreplaceable, and worthy of preservation for future generations (NHPA, 16 USC 470,

Table 9.1 Selection of laws, regulations, and guidelines comprising the cultural resources regulatory framework in the United States

Year[1]	Title and Citation
Prior to first human lunar landing	
1906	American Antiquities Act (16 USC 431–433)
1916	National Park Service Organic Act (16 USC 1–4, 22, 43)
1935	Historic Sites, Buildings, Objects, and Antiquities Act (16 USC 461–467)
1948	Theft of Government Property (18 USC 641)
1949	Federal Property and Administrative Services Act (40 USC 484(k)(3) and (4))
1949	National Trust for Historic Preservation (16 USC 468)
1954	Preservation of American Antiquities (43 CFR Part 3)
1955	Museum Properties Management Act (16 USC 18)
1960	Reservoir Salvage Act
1966	American Battlefield Protection Act (16 USC 469 k)
1966	National Historic Preservation Act (16 USC 470 et seq.)
1969	National Environmental Policy Act (42 USC 4321, and 4331–4335)
After first human lunar landing	
1970	1970 UNESCO Convention on the Means of Prohibiting and Preventing the Illicit Import, Export, and Transfer of Ownership of Cultural Property (19 USC 2601)
1971	Executive Order No. 11593 Protection and Enhancement of the Cultural Environment
1972	Coastal Zone Management Act (16 USC 1451–1456)
1972	National Marine Sanctuaries Act (16 USC 1431–1445) (formerly Marine Protection, Research, and Sanctuaries Act)
1974	Archeological and Historic Preservation Act (16 USC 469-469c-2)
1976	Mining in the National Parks Act (Section 9) (16 USC 1908)
1976	Public Buildings Cooperative Use Act (40 USC 601(a))
1977	Determinations of Eligibility for Inclusion in the National Register (36 CFR Part 63)
1978	American Indian Religious Freedom Act (42 USC 1996 and 1996a)
1978	Secretary of the Interior's Standards for Historic Preservation Projects (36 CFR Part 68)
1979	Archaeological Resources Protection Act (16 USC 470aa–mm)
1981	National Register of Historic Places (36 CFR Part 60)
1983	National Historic Landmarks Program (36 CFR Part 65)
1983	Secretary of the Interior's Standards and Guidelines for Archaeology and Historic Preservation (36 CFR Part 61)
1984	Protection of Archeological Resources (43 CFR Part 7)
1986	Historic Preservation Certifications Pursuant to Section 48(g) and Section 170(h) of The Internal Revenue Code (36 CFR Part 67)
1987	Abandoned Shipwreck Act (43 USC 2101–2106)
1989	Abandoned Shipwreck Act Guidelines (54 FR 13642)
1990	Curation of Federally-Owned and Administered Archeological Collections (36 CFR Part 79)
1990	Internal Revenue Code (Rehabilitation Credit) (26 USC 47)
1990	Native American Graves Protection and Repatriation Act (25 USC 3001)
1995	Native American Graves Protection and Repatriation Act: Final Rule (43 CFR Part 10)

(continued)

Table 9.1 (continued)

Year[1]	Title and Citation
1995	Secretary of the Interior's Standards for the Treatment of Historic Properties (36 CFR Part 68)
1996	Executive Order No. 13006 Locating Federal Facilities on Historic Properties in our Nation's Central Cities
1996	Executive Order No. 13007 Indian Sacred Sites
1999	Procedures for State, Tribal, and Local Government Historic Preservation Programs (36 CFR Part 61)
2000	National Historic Lighthouse Preservation Act (16 U.S.C. § 470w-7)
2000	Protection of Historic Properties (36 CFR Part 800)
2003	Executive Order No. 13287 Preserve America
2004	Sunken Military Craft Act (10 USC 113)

[1]Initial year of passage of act or issuance of relevant guidelines, exclusive of subsequent revisions or amendments

Section 1[b]). Congress also recognized that many of these important pieces of our history are under the ownership of the federal government, and it is the responsibility of the federal government to become stewards of this history so that future generations can benefit from its preservation (16 USC 470-1). As such, Congress directed the Secretary of the Interior to expand and maintain a National Register of Historic Places (NRHP) to afford an additional level of management and consideration for historic properties that represent important parts of our Nation's history and heritage. While Congress may not have had non-earth bound space exploration sites in mind when they passed the NHPA in 1966, Congress did not specifically limit it to sites located on Earth. [In 1984, the National Park Service published Man in Space: A National Historic Landmark Study (Butowsky 1984)].

The NHPA is currently composed of earth-bound districts, sites, buildings, structures, and objects significant in American history, architecture, archaeology, engineering, and culture (16 USC 470a[a]). Eligibility is based on a set of four criteria: association with important events in American history (Criterion A); association with important people in American history (Criterion B); representation of the work of a specific type, or master, or is otherwise architecturally distinctive (Criterion C); or has the potential to provide important information in history (Criterion D) that is underrepresented elsewhere. Noteworthy is that *location* of the resource is not one of the criteria, and is not a condition of the significance. In fact, the resource need not be physically located within the boundaries of the United States to be eligible for, or included in, the NRHP. As one example, the World War II Peleliu Battlefield in Palau, a sovereign island nation in the South Pacific, is listed on the NRHP even though it lies beyond the territorial reach of the United States [NPS 1985 (85001754 NHLS, record number 432280 NRHP)].

Likewise, *age* of the resource is not a condition for listing. Under an exemption in the NRHP that allows for Historic Properties that have achieved significance within the past 50 years, but they must be of "exceptional significance (National Register Bulletin). Many of the Historic Places of the Civil Rights Movement are listed under this exemption. An example is the Brown Chapel AME Church in Selma, Alabama, where on Sunday morning March 7, 1965 (known as Bloody Sunday), about 600 African-American protestors gathered outside Brown Chapel to March from Selma to the state capital in Montgomery. This church was listed on the NRHP in 1982, when it was only 17 years separated from the moment in history that defined its national significance [NPS 1982 (820022009 NRIS, record number 387430 NRHP)]. Accordingly, a Historic Property listed on the NRHP that is neither ancient, nor located on Earth, is not out of the question.

Most properties are evaluated for eligibility for inclusion in the NRHP and only a fraction of the properties and sites actually achieve listing. Although there are no known statistics for properties determined eligible for the NRHP that do not reach listing, in 2013 the National Park Service reported 88,441 total NRHP listings and 1,677,773 total contributing resources (NPS 2013a). Of these, there is an even smaller subset of NRHP properties that rise above the rest as being exceptionally important to American history and heritage. These are eligible for another level of recognition, through the National Historic Landmark Program (NHL). All NHLs are also on the NRHP, but not all NRHP sites are NHLs. As of 2013, there are fewer than 2,500 properties that are designated NHLs (NPS 2013b). Some of these are related to space history, including sites at Cape Canaveral in Florida and Mission Control in Houston.

In the US, only those that are designated NHLs can eventually become a World Heritage site (ICOMOS 2011), though consensus among a national committee that selects from a pool of highly competitive properties. Examples of World Heritage sites include Chaco Canyon (UNESCO Ref: 353rev), the Great Wall of China (UNESCO Ref: 438), Independence Hall (UNESCO Ref: 78), the Acropolis (UNESCO Ref: 404), and the first human footprints by *Australopithecus afarensis* at Laetoli in eastern Africa (Ngorongoro Conservation Area; UNESCO Ref: 39bis). Thus, it can be reasonably argued that the presence of the first human footprints on Earth on the World Heritage List established a precedent for the same recognition being afforded to the first human footprints on another celestial body. The Apollo 11 lunar landing site at Tranquility Base on the moon rises to the same level of significance.

Origins and Goals of the World Heritage List

In the 1950s, international concern for the preservation of universally important, but unprotected, sites was raised following the decision to build the Aswan High Dam in Egypt (UNESCO 2013a). Construction of the dam was to flood the valley containing the Abu Simbel temples, an important archaeological area that includes the Temples of Ramses II (UNESCO 2013b). After appeals from the governments of Egypt and Sudan, the United Nations Educational, Scientific, and Cultural Organization (UNESCO) launched an international safeguarding campaign, costing US$80 million, half of which was donated by 50 countries. The campaign funded archaeological research in the areas to be flooded, and the temples were dismantled, moved to dry ground, and reassembled. Subsequently, similar internationally-funded campaigns were launched in order to save Venice and its lagoon in Italy, the archaeological ruins at Mohenjo-Daro (Pakistan), and restoring the Borobudur Temple Compounds in Indonesia (UNESCO 2013a). Simultaneously, there was a separate movement to conserve important natural landscapes.

Recognizing the international significance of both cultural and natural sites, and the fact that many of the threatened sites are under the jurisdiction of nations (States) that do not have the economic resources to afford them adequate protection, UNESCO sought the help of the International Council on Monuments and Sites (ICOMOS). The result was a convention on the protection of cultural heritage and natural landscapes.

The 1972 Convention Concerning the Protection of the World Cultural and Natural Heritage (Convention) (UNESCO 2013a) forged the concepts of nature conservation and preservation of cultural properties in a single document. The Convention formally recognized the way in which people interact with nature and the fundamental need to preserve the balance between the two (UNESCO 2013a). Among other processes, it provides for a mechanism to formally list such universally important sites on a World Heritage List, and provides for funding to aid in the preservation of listed sites. The 1972 World Heritage Convention (UNESCO 2013a) affirmed the following:

The General Conference of the United Nations Educational, Scientific and Cultural Organization meeting in Paris from 17 October to 21 November 1972, at its seventeenth session,

Noting that the cultural heritage and the natural heritage are increasingly threatened with destruction not only by the traditional causes of decay, but also by changing social and economic conditions which aggravate the situation with even more formidable phenomena of damage or destruction,

Considering that deterioration or disappearance of any item of the cultural or natural heritage constitutes a harmful impoverishment of the heritage of all the nations of the world,

Considering that protection of this heritage at the national level often remains incomplete because of the scale of the resources which it requires and of the insufficient economic, scientific, and technological resources of the country where the property to be protected is situated,

Recalling that the Constitution of the Organization provides that it will maintain, increase, and diffuse knowledge, by assuring the conservation and protection of the world's heritage, and recommending to the nations concerned the necessary international conventions,

Considering that the existing international conventions, recommendations and resolutions concerning cultural and natural property demonstrate the importance, for all the peoples of the world, of safeguarding this unique and irreplaceable property, to whatever people it may belong,

Considering that parts of the cultural or natural heritage are of outstanding interest and therefore need to be preserved as part of the world heritage of mankind as a whole,

Considering that, in view of the magnitude and gravity of the new dangers threatening them, it is incumbent on the international community as a whole to participate in the protection of the cultural and natural heritage of outstanding universal value, by the granting of collective assistance which, although not taking the place of action by the State concerned, will serve as an efficient complement thereto,

Considering that it is essential for this purpose to adopt new provisions in the form of a convention establishing an effective system of collective protection of the cultural and natural heritage of outstanding universal value, organized on a permanent basis and in accordance with modern scientific methods,

Having decided, at its sixteenth session, that this question should be made the subject of an international convention,

Adopts this sixteenth day of November 1972 this Convention.

The Convention was signed first by the United States in 1973, and by 2013, 190 States (nations) had become parties (States Parties) to the Convention (Table 9.2).

Under the Convention, each State that is a party to the Convention may nominate up to two properties per year from its Tentative List of World Heritage properties. Sites proposed for inclusion in the World Heritage List must exhibit Outstanding Universal Value. Section 50 of the Operational Guidelines (World Heritage Centre 2012) defines Outstanding Universal Value as "cultural and/or natural significance which is so exceptional as to transcend national boundaries and to be of common importance for present and future generations of all humanity. As such, the permanent protection of this heritage is of the highest importance to the international community as a whole." Outstanding Universal Value is further qualified through the application of six selection criteria for cultural properties. A World Heritage List cultural site must meet at least one of the following criteria (UNESCO 2013d):

(i)   *represents a masterpiece of human creative genius;*
(ii) *exhibits an important interchange of human values, over a span of time or within a cultural area of the world, on developments in architecture or technology, monumental arts, town-planning or landscape design;*
(iii) *bears a unique or at least exceptional testimony to a cultural tradition or to a civilization which is living or which has disappeared;*
(iv) *is an outstanding example of a type of building, architectural or technological ensemble or landscape which illustrates (a) significant stage(s) in human history;*
(v)   *is an outstanding example of a traditional human settlement, land-use, or sea-use which is representative of a culture (or cultures), or human interaction with the environment especially when it has become vulnerable under the impact of irreversible change; or*
(vi) *is directly or tangibly associated with events or living traditions, with ideas, or with beliefs, with artistic and literary works of outstanding universal significance.*

Four additional criteria apply to natural World Heritage sites. For all proposed sites, the protection, management, authenticity, and integrity of the property are important considerations.

Table 9.2 The list of states to the convention (adapted from UNESCO 2013c)

State party	Date (D/M/Y)	Instrument[1]
United States of America	07/12/1973	R
Iraq	05/03/1974	Ac
Sudan	06/06/1974	R
Algeria	24/06/1974	R
Egypt	07/02/1974	R
Bulgaria	07/03/1974	Ac
Australia	22/08/1974	R
Democratic Republic of the Congo	23/09/1974	R
Nigeria	23/10/1974	Ac
Niger	23/12/1974	Ac
Iran (Islamic Republic of)	26/02/1975	R
Ghana	04/07/1975	R
Jordan	05/05/1975	Ac
Ecuador	16/06/1975	Ac
France	27/06/1975	Ac
Syrian Arab Republic	13/08/1975	Ac
Cyprus	14/08/1975	Ac
Switzerland	17/09/1975	R
Tunisia	10/03/1975	R
Morocco	28/10/1975	R
Senegal	13/02/1976	R
Bolivia (Plurinational State of)	04/10/1976	R
Poland	29/06/1976	R
Canada	23/07/1976	Ac
Pakistan	23/07/1976	R
Germany	23/08/1976	R
Brazil	01/09/1977	Ac
Tanzania, United Republic of	02/08/1977	R
Mali	05/04/1977	Ac
Ethiopia	06/07/1977	R
Guyana	20/06/1977	Ac
Costa Rica	23/08/1977	R
India	14/11/1977	R
Norway	12/05/1977	R
Panama	03/03/1978	R
Nepal	20/06/1978	Ac
Italy	23/06/1978	R
Saudi Arabia	07/08/1978	Ac
Monaco	07/11/1978	R
Argentina	23/08/1978	Ac
Libya	13/10/1978	R

(continued)

Table 9.2 (continued)

State party	Date (D/M/Y)	Instrument[1]
Malta	14/11/1978	Ac
Guatemala	16/01/1979	R
Guinea	18/03/1979	R
Afghanistan	20/03/1979	R
Denmark	25/07/1979	R
Honduras	08/06/1979	R
Nicaragua	17/12/1979	Ac
Haiti	18/01/1980	R
Chile	20/02/1980	R
Sri Lanka	06/06/1980	Ac
Yemen	07/10/1980	R
Seychelles	09/04/1980	Ac
Portugal	30/09/1980	R
Central African Republic	22/12/1980	R
Mauritania	02/03/1981	R
Cuba	24/03/1981	R
Oman	06/10/1981	Ac
Greece	17/07/1981	R
Côte d'Ivoire	09/01/1981	R
Peru	24/02/1982	R
Spain	04/05/1982	Ac
Malawi	05/01/1982	R
Burundi	19/05/1982	R
Benin	14/06/1982	R
Holy See	07/10/1982	A
Cameroon	07/12/1982	R
Zimbabwe	16/08/1982	R
Mozambique	27/11/1982	R
Antigua and Barbuda	01/11/1983	Ac
Lebanon	03/02/1983	R
Bangladesh	03/08/1983	Ac
Turkey	16/03/1983	R
Colombia	24/05/1983	Ac
Jamaica	14/06/1983	Ac
Madagascar	19/07/1983	R
Luxembourg	28/09/1983	R
Mexico	23/02/1984	Ac
Zambia	04/06/1984	R
United Kingdom of Great	29/05/1984	R
New Zealand	22/11/1984	R
Qatar	12/09/1984	Ac

(continued)

Table 9.2 (continued)

State party	Date (D/M/Y)	Instrument[1]
Sweden	22/01/1985	R
Hungary	15/07/1985	Ac
Philippines	19/09/1985	R
Dominican Republic	12/02/1985	R
China	12/12/1985	R
Maldives	22/05/1986	Ac
Saint Kitts and Nevis	10/07/1986	Ac
Gabon	30/12/1986	R
Gambia	01/07/1987	R
Burkina Faso	02/04/1987	R
Lao People's Democratic Republic	20/03/1987	R
Finland	04/03/1987	R
Thailand	17/09/1987	Ac
Congo	10/12/1987	R
Viet Nam	19/10/1987	Ac
Uganda	20/11/1987	Ac
Paraguay	27/04/1988	R
Cape Verde	28/04/1988	Ac
Malaysia	07/12/1988	R
Korea, Republic of	14/09/1988	Ac
Belarus	12/10/1988	R
Russian Federation	12/10/1988	R
Ukraine	12/10/1988	R
Indonesia	06/07/1989	Ac
Uruguay	09/03/1989	Ac
Albania	10/07/1989	R
Mongolia	02/02/1990	Ac
Romania	16/05/1990	Ac
Belize	06/11/1990	R
Venezuela (Bolivarian Republic of)	30/10/1990	Ac
Fiji	21/11/1990	R
Kenya	05/06/1991	Ac
Bahrain	28/05/1991	R
Angola	07/11/1991	R
El Salvador	08/10/1991	Ac
Ireland	16/09/1991	R
Saint Lucia	14/10/1991	R
San Marino	18/10/1991	R
Cambodia	28/11/1991	Ac
Lithuania	31/03/1992	Ac
Georgia	04/11/1992	S

(continued)

Table 9.2 (continued)

State party	Date (D/M/Y)	Instrument[1]
Slovenia	05/11/1992	S
Croatia	06/07/1992	S
Japan	30/06/1992	Ac
Netherlands	26/08/1992	Ac
Tajikistan	28/08/1992	S
Solomon Islands	10/06/1992	A
Austria	18/12/1992	R
Uzbekistan	13/01/1993	S
Czech Republic	26/03/1993	S
Slovakia	31/03/1993	S
Armenia	05/09/1993	S
Bosnia and Herzegovina	12/07/1993	S
Azerbaijan	16/12/1993	R
Kazakhstan	29/04/1994	Ac
Myanmar	29/04/1994	Ac
Turkmenistan	30/09/1994	S
Kyrgyzstan	03/07/1995	Ac
Dominica	04/04/1995	R
Mauritius	19/09/1995	R
Latvia	10/01/1995	Ac
Estonia	27/10/1995	R
Iceland	19/12/1995	R
Belgium	24/07/1996	R
Andorra	03/01/1997	Ac
The Former Yugoslav Republic of	30/04/1997	S
Papua New Guinea	28/07/1997	Ac
South Africa	10/07/1997	R
Suriname	23/10/1997	Ac
Togo	15/04/1998	Ac
Korea, Democratic People's Republic of	21/07/1998	Ac
Grenada	13/08/1998	Ac
Botswana	23/11/1998	Ac
Israel	06/10/1999	Ac
Chad	23/06/1999	R
Namibia	06/04/2000	Ac
Comoros	27/09/2000	R
Kiribati	12/05/2000	Ac
Rwanda	28/12/2000	Ac
Niue	23/01/2001	Ac
Samoa	28/08/2001	Ac
Bhutan	22/10/2001	R

(continued)

Table 9.2 (continued)

State party	Date (D/M/Y)	Instrument[1]
Eritrea	24/10/2001	Ac
United Arab Emirates	11/05/2001	A
Serbia	11/09/2001	S
Liberia	28/03/2002	Ac
Marshall Islands	24/04/2002	Ac
Kuwait	06/06/2002	R
Vanuatu	13/06/2002	R
Micronesia (Federated States of)	22/07/2002	Ac
Barbados	09/04/2002	Ac
Moldova, Republic of	23/09/2002	R
Palau	11/06/2002	Ac
Saint Vincent and the Grenadines	03/02/2003	R
Lesotho	25/11/2003	Ac
Tonga	30/04/2004	Ac
Trinidad and Tobago	16/02/2005	R
Sierra Leone	07/01/2005	R
Swaziland	30/11/2005	R
Guinea-Bissau	28/01/2006	R
Montenegro	03/06/2006	S
Sao Tome and Principe	25/07/2006	R
Djibouti	30/08/2007	R
Cook Islands	16/01/2009	R
Equatorial Guinea	10/03/2010	R
Palestine	08/12/2011	R
Brunei Darussalam	12/08/2011	R
Singapore	19/06/2012	R

States as of 19 September 2012 (current as of November 2013)
[1]Date of deposit of ratification (*R*), acceptance (*Ac*), accession (*A*) or of the notification of succession (*S*)

The benefits of including properties on the World Heritage List is not just symbolic. The mission of World Heritage includes helping help States Parties safeguard World Heritage properties by providing technical assistance and professional training, providing emergency assistance for World Heritage sites in immediate danger, and supporting States Parties' public awareness-building activities for World Heritage conservation (UNESCO 2013d). The World Heritage Fund is maintained by compulsory and voluntary financial contributions by States and donors and is supported by US$4 million annually to provide international assistance that is consistent with the World Heritage mission. Secondary benefits include increased tourism and attention from the academic communities.

The List

As of November 2013, the World Heritage List includes 981 properties composed of 759 cultural, 193 natural, and 29 mixed properties in 160 nations (UNESCO 2013e). Mesa Verde National Park, Yellowstone National Park in the US, the City of Quito, and the Galapagos Islands, in Ecuador, were among the first properties to be listed. More recently, a number of sites were added, including the birthplace of Jesus, the Neolithic site of Çatalhöyük, and Mount Kenya National Park.

However, as of 2013, there are no known 20th century space heritage sites on the World Heritage List. Several World Heritage Sites representing earlier or pre-historic astronomical functions include the Historic Monuments of Deng Feng in "The Centre of Heaven and Earth" and the Jantar Mantar in Jaipur, as well as the Mayan site of Copan and the pre-Hispanic Town of Uxmal, among others.

The World Heritage List also includes sites that are threatened or endangered. Article 11, part 4 of the Convention calls for "the list [to] include only such property forming part of the cultural and natural heritage as is threatened by serious and specific dangers, such as the threat of disappearance caused by accelerated deterioration, large-scale public or private projects or rapid urban or tourist development projects; destruction caused by changes in the use or ownership of the land; major alterations due to unknown causes; abandonment for any reason whatsoever; the outbreak or the threat of an armed conflict; calamities and cataclysms; serious fires, earthquakes, landslides; volcanic eruptions; changes in water level, floods and tidal waves" (UNESCO 2013a). Currently, there are 44 properties that the World Heritage Committee has decided to include on the List of World Heritage in Danger, including the ancient villages of Northern Syria, the minaret and archaeological remains of Jam in Afghanistan, and Iraq's Samarra Archaeological City, Iraq (UNESCO 2013f). These sites are in danger of damage from external forces, much like Tranquility Base and our space heritage on Earth see threats from imminent human visitation and vandalism or neglect, respectively.

Of the 759 cultural sites, 14 are classified as transboundary or a transnational property, which means the property is located on the territory of two or more States parties. One of those transboundary properties is also identified as cultural landscape, the Fertö/Neusiedlersee Cultural Landscape that spans portions of Austria and Hungary.

According to Section 47 of the World Heritage Operational Guidelines (World Heritage Centre 2012), cultural landscapes are "cultural properties and represent the 'combined works of nature and of man' designated in Article 1 of the Convention. They are illustrative of the evolution of human society and settlement over time, under the influence of the physical constraints and/or opportunities presented by their natural environment and of successive social, economic and cultural forces, both external and internal."

In the US, the National Park Service has addressed the concept of cultural or historic landscapes through the issuance of National Register Bulletins that focus on specific types of historic landscapes (Table 9.3). None pertain specifically to space heritage sites, but some aspects are analogous and applicable to space heritage landscapes.

Table 9.3 National Register Bulletins addressing specific landscapes

Bulletin number	Bulletin title	Remarks
n/a	Historic residential suburbs: guidelines for evaluation and documentation for the national register of historic places	Analogous to an historic district, composed of multiple properties
40	Guidelines for identifying, evaluating, and registering America's historic battlefields	Analogous to an archaeological district, which may or may not be composed of elements or features
18	How to evaluate and nominate designed historic landscapes	Applies to landscape architecture (ornamental gardens)
30	Guidelines for evaluating and documenting rural historic landscapes	Geographical areas of land use patterns

Of the National Register Bulletins, number 30 is most applicable to a space heritage cultural landscape. It states, "for the purposes of the [NRHP], a rural historic landscape is defined as a geographical area that historically has been used by people, or shaped or modified by human activity, occupancy, or intervention, and that possesses a significant concentration, linkage, or continuity of areas of land use, vegetation, buildings and structures, roads and waterways, and natural features" (McClelland et al. 1989).

Yet, none of the historic landscapes addressed by National Register bulletins allow for discontiguous landscapes, such as would be the case with space heritage sites. However, a mechanism within the NRHP allows for such an approach: the nomination of multiple properties (Lee and McClelland 1991), such as all properties of a type in a county. It would appear that historic preservation of Tranquility Base and other forms of 20th century space heritage don't fit well within the current legal framework.

Tranquility Base: On the Fringe of Historic Preservation Law

There is a growing consensus over the importance of the historical significance of the first human lunar landing by Neil Armstrong and Buzz Aldrin on July 20, 1969, during the Apollo 11 mission. The historic archaeological site left behind on that day, which is composed of a surface scatter of approximately 106 objects deposited over an area the size of a baseball diamond (Gibson 2001; Westwood et al. 2010; Westwood and O'Leary 2012; Lunar Legacy Project 2000), represents the first time *Homo sapiens* stepped foot on another celestial body. As such, it represents one of the most important technological advancements in human history.

By any reasonable definition, it is a cultural resource—it is composed of a collection of in situ objects that were made, modified, and moved by humans—yet it lacks the antiquity that the average person finds prerequisite. In fact, unlike most archaeological sites, many people remember the site formation process that led to what NASA first called "Tranquility Base." However, as discussed above, even Congress, in establishing the NRHP, recognized that sites do not need to be older in order to be considered significant.

There is compelling evidence to support the notion that Tranquility Base is an NRHP-eligible Historic Property, because: (1) it meets at least several of the eligibility criteria for inclusion in the NRHP; (2) it retains integrity as an untouched, in situ historic-era archaeological site; and (3) is eligible for the exemption for properties that have achieved significance within the past 50 years.

In unprecedented moves by the states of California and New Mexico, it has already been listed on the California Register of Historical Resources (Westwood et al. 2010) and the New Mexico State Register of Cultural Properties (O'Leary et al. 2010). The site is eligible for inclusion in the NRHP but has not yet been formally listed as such. Listing on the NRHP would allow the site to be further designated an NHL and considered for inscription onto the World Heritage List.

Based on the existing federal preservation framework, there are two pathways to achieve designation as an NHL. The first is by way of the traditional process, where a Theme Study is carried out to establish the cultural context and significance, and the final decision to formally designate a property falls to the Secretary of the Interior. Tranquility Base is briefly mentioned in the Man in Space Theme Study (Butowsky 1984), which afforded a number of Earth-bound space heritage sites like Cape Canaveral to be designated NHLs. The alternate route is via Congress, which can designate an NHL through an Act, even if the site has not already been listed on the NRHP. Once a property is an NHL, it will compete with dozens of other potential candidates in the US and could eventually be brought by the US to the Tentative List for eventual inscription onto the World Heritage List by UNESCO.

No matter which route to NHL designation ultimately prevails, there is a sense of urgency here. While it may seem implausible to think that Tranquility Base—a site located over a quarter of a million miles from Earth—is threatened by impact, note that commercial space travel is closer to reality than many people realize. Fueled by abandonment of plans to return to the Moon by NASA, and its reliance on commercial companies to transport to and from the International Space Station (or whatever orbiting lab that replaces it), we are now in the middle of a new "space race"—one that NASA has recently tried to temper through issuance of its Guidelines (NASA 2011).

The next round of threats are very real for site disturbance by well-intended visitors, who may be less aware of the destruction they could do while roaming around the site. We have seen damage to archaeological sites in Chaco Canyon, Mesa Verde, civil war battlefields, and countless other sites by visitors who seek a piece of the history for themselves. In so many instances here on earth, the damage to the sites occurred *before* appropriate cultural resources management practices could be implemented. In the case of Tranquility Base, no humans have since returned to the site, and therefore, it represents a truly in situ archaeological site that has not yet seen cultural damage by humans. It is exceedingly rare to have an archaeological site, so important to human history, remain untouched. Therefore, establishing a legal framework for cultural resources management of lunar heritage sites before humans return to the Moon is crucial. However, in doing so, there are international laws and treaties that must be considered.

A number of international treaties and agreements speak to the "peaceful use of outer space" and while none of them directly address historic preservation, one theme rises to the forefront: like the Earth's oceans, the Moon belongs to all of mankind, but nations who deposit objects onto its surface still retain ownership of those objects. Accordingly to Article VIII of the 1967 Outer Space Treaty (United Nations 1967) and Article 11 of the 1979 United Nations Moon Agreement (United Nations 1979, 2002), no one nation can lay claim to the Moon, either its surface or subsurface (United Nations 1979, 2002), but any nation that transports equipment to the Moon (via humans or remotely), retains ownership of that material and equipment in perpetuity (United Nations 1967). NASA retains ownership to all objects at Tranquility Base, and acknowledges such to this day. With this in mind, the notion that the US could bring to the World Heritage List a surface archaeological site composed of American-owned objects, but not including the lunar surface, is not unreasonable.

In order for Tranquility Base to be considered for inscription on the World Heritage List, the questions regarding compatibility with international treaty must be addressed in the global, political arena. Those legal issues aside (and only acknowledged herein), Tranquility Base must also exhibit Outstanding Universal Value to qualify for consideration.

Outstanding Universal Value of Tranquility Base

The growing body of published literature on the matter (e.g., Westwood and O'Leary 2012; Darrin and O'Leary 2009), has already established that Tranquility Base, the archaeological site formed by Neil Armstrong and Edwin "Buzz" Aldrin on July 20, 1969 qualifies as an historic property that is eligible for inclusion in the NRHP. The site is currently listed on the California Register of Historic Places and the New Mexico State Register of Cultural Properties, for those states' association to the historic first human landing on the Moon. However, the significance of this site extends beyond California, New Mexico, and the US. Tranquility Base also exhibits Outstanding Universal Value. Of the six criteria for inclusion in the World Heritage List, a reasonable case can be argued that Tranquility Base satisfies five:

- Criterion (i) requires that the site represent a masterpiece of human creative genius. Tranquility Base represents a masterpiece of human creative genius in the realm of thought, philosophy, and technology that materialized through the research and development of technology that drew from many nations on Earth.
- Criterion (ii) requires that the site represent an interchange of values over a span of time or within a culture, on developments in architecture and technology, monumental arts, town planning, or landscape design.
  Tranquility Base represents the cumulative interchange of scientific and world cultural values over the span of many decades.

- Criterion (iii) requires that the site bear a unique or exceptional testimony to a cultural tradition or to a civilization, which is living or has disappeared. Space travel and exploration bear both unique and exceptional testimony to the scientific ingenuity of 18th, 19th, and 20th century civilizations.
  Artifacts at Tranquility Base represent many nations on Earth, including a silicon disc carrying statements from Presidents Nixon, Johnson, Kennedy, Eisenhower, and from leaders of 73 other nations.
- Criterion (iv) requires that the site be an outstanding example of a type of building, architectural or technological ensemble or landscape, which illustrates significant stages in history.
  Tranquility Base is the product and result of extraordinary technology that illustrates a significant stage in human history. Artifacts represent mid-20th century technology, and have quickly become technologically obsolete. Related structures and landscapes located on Earth also exhibit extraordinary 20th architecture and technology.
- Criterion (vi) requires that the site be directly tangible or associated with an event or living traditions, with ideas, or with beliefs, with artistic and literary works of outstanding universal significance.
  Tranquility Base represents an event—the first humans to walk upon the Moon—that occurred within the context of the Cold War-era space race, a period of time that possessed its own ideology and worldview. Space technology is the result of ideas and beliefs of the 20th century. Space age ideas have influenced art and literature from the 20th century and continues to influence them in the 21st century.

Despite the Outstanding Universal Value of Tranquility Base, it lies just outside the reach of the World Heritage List, and as such, is vulnerable. The primary reasons for this have to do with the current language in the Convention, the presence of international treaties, and politics. In setting aside the distractions of the political and legal debates, the concept of Tranquility Base as the anchor of a 20th century space heritage cultural landscape can be considered.

Tranquility Base as an Element of a Cultural Landscape

Introduced above was the concept of a cultural landscape, which represents the Combined Works of Nature and Man, and illustrates the evolution of human society and settlement over time, under the influence of the physical constraints and/or opportunities presented by the natural environment. On one level, the artifacts and structures at Tranquility Base are part of a lunar cultural landscape, that takes the shape and form it does because of the interaction between humans and nature. Some of the artifacts left behind at Tranquility Base—such as Moon boots, portable life support systems, and even human waste collection devices—were specifically engineered and manufactured to allow the human body to survive or function in the unfriendly environment of space. The distribution of artifacts and features, like footpaths, are specifically related to the Apollo crew's mission to navigate, explore, and sample the lunar surface.

However, it is also reasonable to look at space heritage on a broader level—one that includes space heritage sites on Earth. In the simplest of terms, Tranquility Base would not exist were it not for Cape Canaveral, Mission Control, and dozens of other launch, communication, tracking, research and development, and manufacturing facilities and sites on Earth. The locations of some of these facilities, like Cape Canaveral, were reportedly selected to accommodate orbit entry and maximize efficiency during liftoff. The launch stands were specifically designed to accommodate the thrust necessary to hurl rockets beyond the Earth's gravitational control. In one sense, all of the facilities on Earth related to the Apollo missions and human space exploration in general were designed, functioned, or sited for specific reasons that relate to the interaction between humans and the environment, both on Earth and beyond.

However, not all facilities on Earth contribute equally to the significance of Tranquility Base in the same way. The launch facilities at Cape Canaveral contribute more to the historical significance than the bolt factory that made a small piece of equipment on the *Eagle* lander. As important as that piece of equipment was to the success of the mission, its *historical* significance is negligible. Therefore, when considering a potential space heritage cultural landscape, it is important to identify those elements that contribute to the *historical* significance of the first human lunar landing.

Proposed Human Lunar Exploration Heritage Cultural Landscape

Given the outstanding universal value of Tranquility Base, and the related and historically significant facilities on Earth, a transnational serial nomination of a cultural landscape that includes, but is not limited to, Tranquility Base, seems appropriate. Establishing such a human space exploration heritage cultural landscape requires careful consideration of the geographical and temporal extent, and the nations and facilities involved. With these parameters defined, the cultural landscape can be proposed for inscription on the World Heritage List.

Establishing a Transnational Serial Nomination

Under Sections 137 through 139 of the World Heritage Operational Guidelines (World Heritage Centre 2012), a transnational serial property is composed of two or more component parts, related by three clearly defined links:

(a) Component parts should reflect cultural, social or functional links over time that provide, where relevant, landscape, ecological, evolutionary or habitat connectivity.
(b) Each component part should contribute to the Outstanding Universal Value of the property as a whole in a substantial, scientific, readily defined and

discernible way, and may include, inter alia, intangible attributes. The resulting Outstanding Universal Value should be easily understood and communicated.

(c) Consistently, and in order to avoid an excessive fragmentation of component parts, the process of nomination of the property, including the selection of the component parts, should take fully into account the overall manageability and coherence of the property ... and provided it is the series as a whole—and not necessarily the individual parts of it—which are of Outstanding Universal Value.

The Operational Guidelines further specify that a serial nominated property may occur within the territory of different States Parties, which need not be contiguous and is nominated with the consent of all States Parties concerned and may be submitted for evaluation over several nomination cycles, provided that the first property nominated is of Outstanding Universal Value in its own right. Accordingly, a transnational serial nomination could begin with the nomination of Tranquility Base by the US, and over subsequent nomination cycles, additional sites on the Earth and Moon can be advanced by the US and other States Parties and added to the inscription. The identity of those additional inscriptions, which collectively would fill out a cultural landscape, would be defined by their period of significance and association with the history of lunar exploration. The entirety of the cultural landscape need not be defined in full prior to the inscription of one or more elements of the landscape.

Determining the Period of Significance

It can be reasonably argued that human desire for space exploration and reaching the Moon began thousands of years before the first launch vehicle rose into the atmosphere, and that it continues to this day. In an effort to designate a more manageable World Heritage Site and further define the assemblage of contributing elements, it is necessary to focus the period of significance. Reflecting on broader patterns of space technology and events, two paradigm shifts emerge as potential terminal dates for the period of significance: the development of jet assisted take-off beginning in 1936 and the more recent and shift from US government-funded exploration toward a reliance on private-sector, commercial, and international endeavors.

A period of significance that spans nearly three-quarters of a century is feasible and may ultimately be desirable, but that which is more closely associated with physical human presence on the Moon and anchored by Tranquility Base seems like an appropriate place to start. As such, the period of significance can be further bracketed by the first human in space (Cosmonaut Yuri Gagarin on 12 April 1961) through the last Apollo mission (the return of Apollo 17 to Earth on 19 December 1972). Interestingly, it was also in 1961 when President John F. Kennedy issued his directive to Congress to send a man to the Moon and return him home safely by the end of the decade.

Identifying the States Parties Involved

During the period of 1961 through 1972, the roster of space-faring nations was more limited than it is today. The space race was set within the context of the mid-20th century Cold War era, which was heightened by the launch of the first human into space by the Soviet Union. As Cosmonaut Yuri Gagarin climbed into orbit on 12 April 1961, the American public incorrectly assumed that the Soviets were taking a huge lead in the development of the nearly unstoppable Intercontinental Ballistic Missiles (ICBMs). Widespread fear ensued, fueling the US space program (Friedman 2000) and leading President Kennedy to issue his challenge to Congress. Over the next decade, the US and Soviet Union battled for the bragging rights to have one of their citizens be the first human to step foot on the Moon. In retrospect, it would seem that the space race was more about politics than it was about an incredibly important milestone in human history.

Meanwhile, other nations were contributing important infrastructure associated with deep space communication and tracking of the Apollo missions. The Deep Space Network is composed of three communication satellite facilities placed approximately 120° apart around the world. Collectively, these three facilities allow for constant observation of spacecraft as the Earth rotates (NASA 2013).

In California's Mojave Desert, the first of three is referred to as the Goldstone Tracking Station, and the original antenna named after the first spacecraft with which it communicated—Pioneers 3 and 4 (1958–1959). NASA required multiple antennas at Goldstone because of the narrow beam width when Apollo 11 was near the moon—one was focused on the Command and Service Module that was being piloted by Michael Collins, and the other focused on the Descent Module on the lunar surface. The Deep Space Network also includes the Madrid Deep Space Communications Complex, located 65 km west of Madrid, near the town of Robledo de Chavela, Spain. Located 40 km southwest of Canberra, Australia, near the Tidbinbilla Nature Reserve, is the last of the three facilities.

Therefore, the US, former USSR, Australia, and Spain have been identified as having direct ties to Apollo during the period of significance described above. It is highly likely that additional archival research will yield additional nations with ties to lunar exploration between the years of 1961 and 1972. As additional States Parties step forward with such facilities and sites within their jurisdiction, they could be included in the transnational serial nomination under subsequent nomination cycles.

Identifying the Facilities

Based on the position presented above, there are at least 78 sites and facilities on Earth and the Moon that fall within the period of significance of 1961 to 1972 and are affiliated in some manner with human space flight—and more specifically, lunar exploration. These are presented in Table 9.4.

Table 9.4   Potential elements of the cultural landscape

Name	Associated country	Current location	Mission year	Registry status
Ranger 4	US	Moon	1962	–
Ranger 6	US	Moon	1964	–
Ranger 7	US	Moon	1964	–
Luna 5	USSR	Moon	1965	–
Luna 7	USSR	Moon	1965	–
Luna 8	USSR	Moon	1965	–
Ranger 8	US	Moon	1965	–
Ranger 9	US	Moon	1965	–
Luna 10	USSR	Moon	1966	–
Luna 11	USSR	Moon	1966	–
Luna 12	USSR	Moon	1966	–
Luna 13	USSR	Moon	1966	–
Luna 9	USSR	Moon	1966	–
Lunar Orbiter 1	US	Moon	1966	–
Lunar Orbiter 2	US	Moon	1966	–
Lunar Orbiter 3	US	Moon	1966	–
Surveyor 1	US	Moon	1966	–
Surveyor 2	US	Moon	1966	–
Explorer 35 (IMP-E)	US	Moon	1967	–
Lunar Orbiter 4	US	Moon	1967	–
Lunar Orbiter 5	US	Moon	1967	–
Surveyor 3	US	Moon	1967	–
Surveyor 4	US	Moon	1967	–
Surveyor 5	US	Moon	1967	–
Surveyor 6	US	Moon	1967	–
Surveyor 7	US	Moon	1967	–
Luna 14	USSR	Moon	1968	–
Apollo 10 LM (Snoopy) descent stage	US	Moon	1969	–
Apollo 11 LM ascent stage	US	Moon	1969	–
Apollo 11 LM-5 (Eagle) descent stage	US	Moon	1969	CRHR; NMSRCP
Apollo 11 (106 + objects at Tranquility Base)	US	Moon	1969	CRHR; NMSRCP
Apollo 12 LM ascent stage	US	Moon	1969	–
Apollo 12 LM-6 (Intrepid) descent stage	US	Moon	1969	–
Luna 15	USSR	Moon	1969	–
Apollo 13 S-IVB (S-IVB-508)	US	Moon	1970	–
Luna 16 descent stage	USSR	Moon	1970	–

(continued)

Table 9.4 (continued)

Name	Associated country	Current location	Mission year	Registry status
Luna 17 and Lunokhod 1	USSR	Moon	1970	–
Apollo 14 LM-8 (Antares) descent stage	US	Moon	1971	–
Apollo 14 LM-8 ascent stage	US	Moon	1971	–
Apollo 14 S-IVB (S-IVB-509)	US	Moon	1971	–
Apollo 15 LM-10 (Falcon) descent stage	US	Moon	1971	–
Apollo 15 LM-10 ascent (Falcon) stage	US	Moon	1971	–
Apollo 15 Lunar Rover	US	Moon	1971	–
Apollo 15 S-IVB (S-IVB-510)	US	Moon	1971	–
Apollo 15 subsatellite	US	Moon	1971	–
Luna 18	USSR	Moon	1971	–
Luna 19	USSR	Moon	1971	–
Apollo 16 LM-11 (Orion) ascent stage	US	Moon	1972	–
Apollo 16 LM-11 (Orion) descent stage	US	Moon	1972	–
Apollo 16 Lunar Rover	US	Moon	1972	–
Apollo 16 S-IVB (S-IVB-511)	US	Moon	1972	–
Apollo 16 subsatellite	US	Moon	1972	–
Apollo 17 LM-12 (Challenger) ascent stage	US	Moon	1972	–
Apollo 17 LM-12 (Challenger) descent stage	US	Moon	1972	–
Apollo 17 Lunar Rover	US	Moon	1972	–
Apollo 17 S-IVB (S-IVB-512)	US	Moon	1972	–
Luna 20 descent stage	USSR	Moon	1972	–
Jet Propulsion Laboratory	US	California	–	–
Ames Research Center	US	California	–	–
Dryden Flight Center / Edwards Air Force Base	US	California	–	–
White Sands Propulsion Facility	US	New Mexico	–	–
White Sands Test Facility	US	New Mexico	–	–
Pioneer Deep Space Station (DSN)	US	California	–	NHL
Canberra Tracking Station (DSN)	Australia	Canberra	–	–

(continued)

Table 9.4 (continued)

Name	Associated country	Current location	Mission year	Registry status
Madrid Tracking Station (DSN)	Spain	Madrid	–	–
Santa Susana Field Lab	US	California	–	–
Apollo Mission Control Center	US	Texas	–	NHL
Cape Canaveral Air Force Station	US	Florida	–	NHL
Lunar Landing Research Facility	US	Virginia	–	NHL
Saturn V Dynamic Test Stand	US	Alabama	–	NHL
Saturn V Launch Vehicle	US	Alabama	–	NHL
Redstone Test Stand	US	Alabama	–	NHL
Rendezvous Docking Simulator	US	Virginia	–	NHL
Rocket Propulsion Test Complex	US	Mississippi	–	NHL
Space Environment Simulation Laboratory	US	Texas	–	NHL
Space Launch Complex 10	US	California	–	NHL
Twenty-five-foot Space Simulator	US	California	–	NHL
Baikonur Cosmodrome (Tyuratam)	USSR	Kazakhstan	–	–

Conclusions

At the time that the Convention was ratified, properties must be located within the territory of the nominating State in order to be considered for inclusion in the World Heritage List. Nowhere in the Convention or its Operational Guidelines (World Heritage Centre 2012) is there any indication that the authors were thinking of cultural landscapes off of Earth; nowhere does it specifically exclude them. This is likely a reflection of then-current worldview. The thrust behind the establishment of the World Heritage List in the 1950s (the building of the Aswan High Dam in Egypt) and the current events at the time of the issuance of the Operational Guidelines in 1972 (as the Tranquility Base was in existence only 3 years) shaped the scope of the WHL program into one that focused on preservation of important sites on Earth that could be damaged or destroyed by humans.

Allowing space heritage sites off of Earth to be included in the WHL may require changes to the Convention and to its Operational Guidelines. First, a change to the Operational Guidelines would be necessary to address immovable heritage that is likely to become moveable. These types of resources are not currently covered by the Convention. Although many of the contributing elements on Earth are immovable and are likely to remain so, the sites on the Moon are composed of artifacts and objects that could be moved and returned to Earth, should someone so desire in the future.

Second, changes would be required to allow for States Parties—or the World Heritage Committee itself, through consensus—to nominate properties that are currently situated outside of their territories, or in areas that governed by international space treaties. Such a change would similarly address conflicts with significant historical sites within international waters or on Antarctica, for example. Such amendments would need to be compatible with existing international treaties, as well.

Perhaps the most important change is one of perception—that not all sites that exhibit Outstanding Universal Value are beautiful buildings with clearly visible boundaries that lie squarely within the unquestionable territory of the nominating State. Changes to the Convention and its Operational Guidelines are necessary. Neither will be easy, but change is inevitable as our heritage continues to diversify. Our generation has the opportunity and responsibility to do something now to preserve a part of our collective human heritage that lies just beyond our reach, on the fringe of historic preservation.

References

Butowsky, H. A. (1984). *Man in Space: A national historic landmark theme study*. Washington DC. U.S. National Park Service, Man in Space Theme Study. Accessed April 2011. http://www.cr.nps.gov/history/online_books/butowsky3/spacet.htm

Darrin, A. G., & O'Leary, B. L. (Eds.). (2009). *Handbook of space engineering, archaeology and heritage*. Boca Raton: CRC Taylor and Francis Press.

Friedman, N. (2000). *The fifty year war: Conflict and strategy in the Cold War*. Annapolis, MD: Naval Institute Press.

Gibson, R. (2001). *Lunar archaeology: The application of federal historic preservation law to the site where humans first set foot upon the moon*. Master's Thesis, Department of Anthropology, New Mexico State University, Las Cruces. MS on file at Dept.

International Committee on Monuments and Sites (ICOMOS) (2011). *World heritage convention and ICOMOS*. Accessed April 2011. http://www.usicomos.org/heritage.

Lee, A. J., & McClelland, L. F. (1991) (revised 1999). *How to complete the national register multiple property documentation form*. National Register Bulletin 16b. Accessed January 8, 2014. http://www.nps.gov/nr/publications/bulletins/pdfs/nr16b.pdf

Lunar Legacy Project (2000). *Archaeological inventory at tranquility base*. Accessed December 2013. http://spacegrant.nmsu.edu/lunarlegacies/artifactlist.html

McClelland, L. F., Keller, J. T., Keller, G. P., & Melnick, R. Z. (1989) (revised 1999). *Guidelines for evaluating and documenting rural historic landscapes*. National Register Bulletin 30. Accessed January 8, 2014. http://www.cr.nps.gov/nr/publications/bulletins/nrb30/

NASA (2011). *NASA's recommendations to space-faring entities: How to protect and preserve the historic and scientific value of U.S. Government lunar artifacts*. Issued July 30, 2011. Accessed January 2014. http://www.nasa.gov/pdf/617743main_NASA-USG_LUNAR_HISTORIC_SITES_RevA-508.pdf

NASA (2013). *About the deep space network*. Accessed November 2013. http://deepspace.jpl.nasa.gov/dsn/about.html

National Park Service (1982). *Brown Chapel African Methodist Episcopal Church*. Accessed April 2011. http://pdfhost.focus.nps.gov/docs/NRHP/Text/82002009.pdf

National Park Service (1985). *Peleliu battlefield*. National Register of Historic Places. Accessed April 2011. http://pdfhost.focus.nps.gov/docs/NHLS/Text/85001754.pdf

National Park Service (2013a). *National register of historic places*. Accessed November 2013. http://www.nps.gov/NR/

National Park Service (2013b). *National register of historic places*. Accessed November 2013. http://www.nps.gov/nhl/whatis.htm

O'Leary, B., Westwood, L., Gibson, R., Versluis, J., with contributions by Bliss, S., DeBry, R., Punke, M., Sam, D., Slocum, R., & Vela, J. (2010). *New Mexico Register of cultural properties nomination forms for the objects and structures at tranquility base*. Presented to the Historic Preservation Division of the State of New Mexico, Santa Fe. Listed on April 10, 2010.

United Nations (1967). *Treaty on principles governing the activities of states in the exploration and use of outer space, including the Moon and other celestial bodies*. Accessed April 2011. http://www.oosa.unvienna.org/oosa/en/SpaceLaw/gares/html/gares_21_2222.html

United Nations (1979). *Agreement Governing the Activities of States on the Moon and other celestial bodies*. Accessed April 2011. http://www.oosa.unvienna.org/oosa/en/SpaceLaw/gares/html/gares_34_0068.html

United Nations (2002). *United Nations treaties and principles on outer Space: Text of treaties and principles governing the activities of States in the exploration and use of outer space, adopted by the United Nations General Assembly*. United Nations Publication Sales No. E.02.I.20. New York: UN. Accessed April 2011. http://www.oosa.unvienna.org/pdf/publications/STSPACE11E.pdf

UNESCO (2013a). *World heritage convention*. Accessed November 2013. http://whc.unesco.org/en/convention/

UNESCO (2013b). *List of States to the convention*. Accessed November 2013. http://whc.unesco.org/en/statesparties/

UNESCO (2013c). *Criteria*. Accessed November 2013. http://whc.unesco.org/en/criteria/

UNESCO (2013d). *About*. Accessed November 2013. http://whc.unesco.org/en/about/

UNESCO (2013e). *World heritage list*. Accessed November 2013. http://whc.unesco.org/en/list/

UNESCO (2013f). *World heritage in danger*. Accessed November 2013. http://whc.unesco.org/pg.cfm?cid=86

Westwood, L., & O'Leary, B. (2012). The archaeology of tranquility base. *Space Times Magazine, 4*(51).

Westwood, L., Gibson, R., O'Leary, B., & Versluis, J. (2010). *California register of historical resources nomination forms for the objects associated with tranquility base*. Submitted to the California Historical Resources Commission, Sacramento. Listed on January 29, 2010. http://www.ohp.parks.ca.gov/pages/1067/files/tranquility%20base_draft.pdf

World Heritage Centre (2012). *Operational guidelines for the implementation of the world heritage convention*. Intergovernmental Committee for the Protection of the World Cultural and Natural Heritage. Accessed January 2014. http://whc.unesco.org/archive/opguide13-en.pdf

Erratum to: The Space Shuttle *Discovery,* Its Scientific Legacy in a Museum Context

Hanna M. Szczepanowska

Erratum to:
Chapter 5 in: B.L. O'Leary and P.J. Capelotti (eds.),
Archaeology and Heritage of the Human Movement into Space,
DOI 10.1007/978-3-319-07866-3_5

Figure 5.1 caption is updated; hence, it should read as below:

Fig. 5.1 *Enterprise* and *Discovery* facing facing each other on the grounds of NASM's Udvar-Hazy Center, April 17, 2012 (Image 2012, H. Szczepanowska)

The online version of the original chapter can be found under
DOI 10.1007/978-3-319-07866-3_5

H.M. Szczepanowska (✉)
MCI Smithsonian Institution, Suitland, MD 20746, USA
e-mail: hszczepanowska1@gmail.com

E1

© Springer International Publishing Switzerland 2015
B.L. O'Leary and P.J. Capelotti (eds.), *Archaeology and Heritage of the Human Movement into Space*, Space and Society, DOI 10.1007/978-3-319-07866-3_10

Artwork of Fig. 5.4 has been replaced; hence, it should read as below:

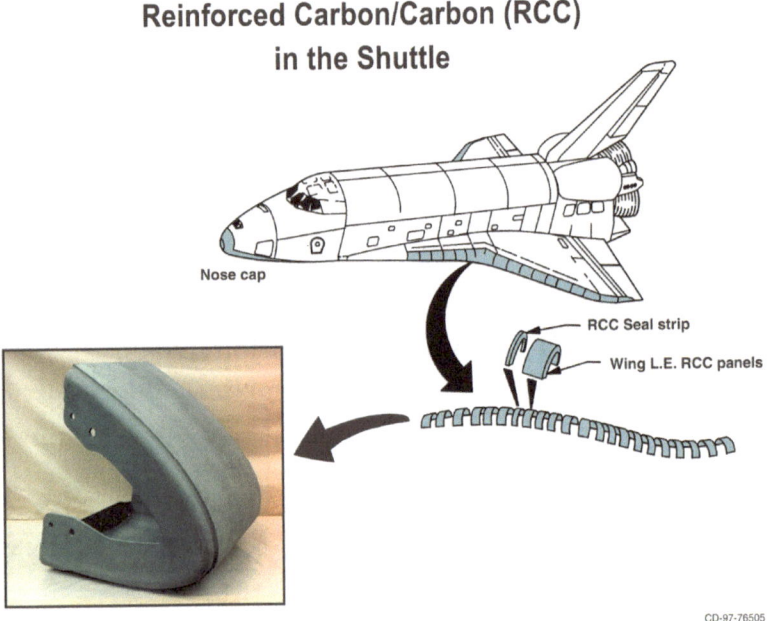

Fig. 5.4 Reinforced carbon–carbon (RCC) shield applied on the leading edges and nose of the shuttle orbiter (courtesy of Dr. Nathan Jacobs, NASA Glenn Research Center)

Artwork of Fig. 5.5 has been replaced; hence, it should read as below:

Fig. 5.5 Detail of the HRSI tile reveals in the chipped areas structure under the black coating, RTV adhesive and remains of an insulation pad to which it was attached (Image 2008, H. Szczepanowska)

Index